Nelson Advanced Science

Genetics, Evolution and Biodiversity

revised edition

John Adds • Erica Larkcom • Ruth Miller

Series Editor: Martin Furness-Smith

Nelson Thornes

First published in 2001 by:
Nelson Thornes Ltd
Delta Place
27 Bath Road
CHELTENHAM
GL53 7TH
United Kingdom

This edition published in 2004.

08 / 10 9 8 7

A catalogue record for this book is available from the British Library

ISBN 978 0 7487 7492 0

Illustrations and page make-up by Hardlines and Wearset Ltd

Printed and bound in Croatia by Zrinski

Acknowledgements
The authors and publisher are grateful for permission to include copyright material supplied by
the following people and organisations:

David Irvine of United Agri Products; English Nature; Field Studies Council.
The examination questions are reproduced by permission of London Examinations, a division of
Edexcel Foundation.

Artwork
Christine Grey-Wilson supplied the sketches for figures 5.1 and 5.2; Dean Madden, National
Centre for Biotechnology Education, Reading University provided references for figures 9.5
and 9.6 and part of figure 9.7 (from NCBE newsletters); figure 9.8 is from 'Food Biotechnology
– an introduction'; John Schollar at the National Centre for Biotechnology Education provided
references for figures 10.10, 10.12, 10.13 and part of figure 9.7; *The Cambridge Encyclopaedia
of Human Evolution* (eds S. Jones, R. Martin and D. Pilbeam) for figures 10.2–9, 10.14, 10.19–20,
10.22, 10.24–27

Photographs
Anne Stephens: 5.6 top, middle, bottom;
Erica Larkcom: 2.5, 2.6 top, bottom, 3.8, 4.16 top, bottom, 5.3, 5.4 left, right, 5.5, 6.25, 10.32,
    11.4, 11.15;
Geoscience Features Picture Library: 2.1;
Getty Images Stone: Art Wolfe 10.1 top, upper middle, Daniel J Cox 10.1 bottom,
    Manoj Shah 10.1 lower middle;
John Bebbington FRPS, Field Studies Council: 5.7;
John Innes Centre: 9.7a, b;
Philip Harris: 3.5;
Planet Earth Pictures: 4.12, 8.3;
Science Photolibrary: 1.9, 6.2a, 6.11 top, bottom, 6.31, J C Revy 6.2b, James
    Holmes/Cellmark Diagnostics 7.5, David Parker 9.11, John Reader 10.10a, b, Phillipe Plailly/
    Eurelios 10.22, Sinclair Stammers 2.7;
Techne (Cambridge) Ltd: 9.10c;
Telegraph Colour Library: Greg Pease cover;
Werner Forman Archive: 10.38.

Every effort has been made to contact copyright holders, and the publishers apologise
for any omissions.

Contents

CONTENTS

CONTENTS

Introduction

This series has been written by an experienced team of Examiners and others involved directly with the Edexcel Advanced Subsidiary (AS) and Advanced (A) GCE Biology and Biology (Human) specification and its assessment.

Genetics, Evolution and Biodiversity is one of four books in the Nelson Advanced Science (NAS) series. These books have been developed to match the requirements of the Edexcel specification, but they will also be useful for other Advanced Subsidiary (AS) and Advanced (A) courses.

Genetics, Evolution and Biodiversity covers Unit 5B and 5H of the Edexcel specification for the Advanced (A) course.

Unit 5B – Genetics, evolution and biodiversity – for the Biology specification, covers:
• photosynthesis, including mineral nutrition
• control of growth in plants
• biodiversity: classification, the distribution of plants and animals, succession, populations and conservation
• genetics and evolution.

Unit 5H – Genetics, human evolution and biodiversity – for the Biology (Human) specification, covers:
• genetics and evolution
• human evolution
• human populations
• biodiversity: distribution of plants and animals, succession, control of insect pests and conservation.

An outline of the assessment of this unit, including the synoptic element, can be found in the introduction to the Assessment questions on page 169.

It should be noted that there are topics common to both specifications. Students of Biology and Biology (Human) require an understanding of variation, genes and alleles, sources of new inherited variation and gene technology, together with some knowledge of classification, studies on the distribution of plants and animals and selected methods of conservation.

In both specifications, there are important links with the Advanced Subsidiary (AS) course. For example, a knowledge of the structure of DNA, chromosomes and the genetic code from Unit 1 is developed and extended in genetics and evolution. An understanding of ecosystems is extended and applied to the distribution of plants and animals and a knowledge of human influences on the environment is central to the appreciation of methods of conservation.

Unit 6 – Synoptic and practical assessment. There is no specification content for this unit, although there are two assessment components.

In these components, students either:
• carry out and submit a report on an individual study and sit a synoptic paper **or**
• take a written alternative test (W2) and sit a synoptic paper.

The other student books in the series are:
• *Molecules and Cells* covering Unit 1 and containing the Appendix, which provides the physical science background for the complete course
• *Exchange and Transport, Energy and Ecosystems* covering Units 2B, 2H and 3
• *Respiration and Coordination* covering Unit 4 and including the Options.

INTRODUCTION

In *Molecules and Cells*, there is an **Appendix**, which provides the **physical science background** that you need in the study of the Biology and Biology (Human) specifications. Much of the information in the appendix is particularly relevant to the topics covered in Unit 1, but it is a useful reminder of some basic scientific concepts that may need to be referred to throughout the course.

Other resources in this series

NAS *Tools, Techniques and Assessment in Biology* is a course guide for students and teachers. For use alongside the four student texts, it offers ideas and support for practical work, fieldwork and statistics. Key Skills opportunities are identified throughout. This course guide also provides advice on the preparation for assessment tests (examinations).

NAS *Make the Grade in AS Biology with Human Biology* and *Make the Grade in A2 Biology with Human Biology* are Revision Guides for students and can be used in conjunction with the other books in this series. They help students to develop strategies for learning and revision, to check their knowledge and understanding, and to practise the skills required for tackling assessment questions.

Features used in this book – notes to students

The NAS Biology student books are particularly written to help you understand and learn the information provided, and to help you to apply this information to your coursework:

The **text** offers complete and self-contained coverage of all the topics in each Unit. Key words are indicated in **bold**. The headings for sub-sections have been chosen to link with the wording of the specification wherever possible.

In the margins of the pages, you will find:
- **definition boxes** where key terms are defined. These reinforce and sometimes expand definitions of key terms used in the text.
- **questions** to test your understanding of the topics as you study them. Sometimes these questions take the topic a little further and stimulate you to think about how your knowledge can be applied.

Included in the text are boxes with:
- **background information** designed to provide material which could be helpful in improving your understanding of a topic. This material could provide a link between knowledge gained from GCSE and what you are required to know for AS and A GCE. It could be more information about a related topic or a reminder of material studied at a different level.
- **additional** or **extension** material which takes the topic further. This material is not strictly part of the Edexcel specification and you will not be examined on it, but it can help you to gain a deeper understanding, extending your knowledge of the topic.

In the specification, reference is made to the ability to recognise and identify the general formulae and structure of biological molecules. You will see that we have included the structural chemical formulae of many compounds where we think that this is helpful in gaining an understanding of the composition of the molecules and appreciating how bond formation between monomers results in the formation of polymers. It should be understood that you will not be expected to memorise or reproduce these structural chemical formulae, but you should be able to recognise and reproduce the general formulae for all the molecules specified.

The chapters correspond to the sections of the specification. At the end of each chapter you will find the **practical investigations** linked to the topics covered. These practical investigations are part of the specification and you could be asked questions on them in the Unit tests. Each practical has an introduction, putting it into the context of the topic, and sufficient information about materials and procedure to enable you to carry out the investigation. In addition, there are suggestions as to how you should present your results and questions to help you with the discussion of your findings. In some cases, there are suggestions as to how you could extend the investigation so that it would be suitable as an individual study.

At the end of the book, there are **assessment questions**. These have been selected from past examination papers and chosen to give you as wide a range of different types of questions as possible. These should enable you to become familiar with the format of the Unit Tests and help you to develop the skills required in the examination. **Mark schemes** for these questions are provided so that you can check your answers and assess your understanding of each topic.

Note to teachers on safety

When practical instructions have been given, we have attempted to indicate hazardous substances and operations by using standard symbols and appropriate precautions. Nevertheless, teachers should be aware of their obligations under the Health and Safety at Work Act, Control of Substances Hazardous to Health (COSHH) Regulations, and the Management of Health and Safety at Work Regulations. In this respect they should follow the requirements of their employers at all times. In particular, they should consult their employer's risk assessments (usually model risk assessments in a standard safety publication) before carrying out any hazardous procedure or using hazardous substances or microorganisms.

In carrying out practical work, students should be encouraged to carry out their own risk assessments, that is, they should identify hazards and suitable ways of reducing the risks from them. However, they must be checked by the teacher. Students must also know what to do in an emergency, such as a fire.

Teachers should be familiar and up to date with current advice on safety, which is available from professional bodies.

Teachers are strongly advised to refer to Safety Codes of Practice and Guidelines produced by Education Authorities or by Governing Bodies of schools and colleges. This is particularly important in practical work on students, such as investigations into the effects of exercise, in which students should be sufficiently fit and willing to participate. Some practical activities, in particular those involving measurement of body fat and comparisons of fitness, should be approached with sensitivity and understanding.

Acknowledgements

The authors would like to thank Sue Howarth and David Hartley for their help and support during the production of this book, as well as John Schollar and Dean Madden, The National Centre for Biotechnology Education, The British Nutrition Foundation and Sainsburys plc.

About the Authors

John Adds is Chief Examiner for AS and A GCE Biology and Biology (Human) and Head of Biology at Abbey College, London.

Erica Larkcom is Deputy Director of Science and Plants for Schools at Homerton College, Cambridge, and a former Subject Officer for A level Biology.

Ruth Miller is a former Chief Examiner for AS and A GCE Biology and Biology (Human) and former Head of Biology at Sir William Perkins's School, Chertsey.

Unit 5 – What do I need to study?

This table shows you which chapters or parts of chapters you need to study for your chosen Units in the Edexcel specification. Sections of the book relevant only to Biology Unit 5 are indicated by a blue stripe in the margin; sections relevant only to Biology (Human) Unit 5 are indicated by a pink stripe.

Topic	Page no.	Unit 5B*	Unit 5H*
Chapter 1			
Photosynthesis	1	✓	
Chapter 2			
Control of growth in plants	17	✓	
Chapter 3			
Classification	28	✓	
Distribution of plants and animals	31	✓	✓
Qualitative and quantitative field techniques	37	✓	✓
Succession	40	✓	✓
Chapter 4			
Populations and communities	48	✓	
Control of insect populations	52	✓	✓
Chapter 5			
Conservation	63	✓	✓
Chapter 6			
Genes, alleles and sources of new inherited variation	77	✓	✓
Chapter 7			
Genetic counselling	97		✓
Chapter 8			
Environmental change and evolution	106	✓	✓
Chapter 9			
Gene technology	115	✓	✓
Chapter 10			
Human evolution	128		✓
Chapter 11			
Human populations	153		✓

*5B = Biology Unit 5, 5H = Biology (Human) Unit 5

Photosynthesis

All living organisms need energy for growth and maintenance. **Autotrophic** organisms are able to use external sources of energy in the synthesis of their organic food materials, whereas **heterotrophic** organisms must be supplied with ready-made organic compounds from which to derive their energy. Algae, green plants and certain prokaryotes can obtain the energy for synthesis directly from the Sun's radiation. It is then used to build up essential organic compounds from inorganic molecules. Such organisms are called **photosynthetic** and possess special pigments which can absorb the necessary light energy.

A few specialised prokaryotes are able to use energy derived from certain types of chemical reaction in the synthesis of organic molecules from inorganic ones. These organisms are called **chemosynthetic** and include the nitrifying bacteria, *Nitrosomonas* and *Nitrobacter,* which are important in the nitrogen cycle. All other organisms are heterotrophic and are dependent on autotrophic organisms for their energy supplies, so must feed on plants, or on other animals which have eaten plants.

Photosynthesis is the source of energy and organic materials for other organisms besides plants. So it can be seen that the ultimate source of all metabolic energy is the Sun and photosynthesis is responsible for the maintenance of life on Earth.

The process of photosynthesis

Photosynthesis in green plants is the process in which energy from the Sun is transformed into chemical bond energy in organic molecules. It is a process in which energy is transduced from one form to another and results in the inorganic molecules, carbon dioxide and water, being built up into organic molecules. Oxygen is produced as a waste product. In green plants, the first stable organic molecules to be formed in photosynthesis are simple sugars. The general equation for photosynthesis can be written as:

$$CO_2 + 2H_2O \xrightarrow[\text{chlorophyll}]{\text{light}} (CH_2O) + O_2 + H_2O$$

carbon dioxide + water → carbohydrate + oxygen + water

NB: Water is a product as well as a reactant in the process.

Photosynthesis occurs in two stages:
- the **light-dependent stage**, which requires light energy and results in the production of **ATP (adenosine triphosphate)**, **NADPH+H$^+$ (reduced nicotinamide adenine dinucleotide phosphate)** and oxygen (as a waste product).
- the **light-independent stage** in which the NADPH+H$^+$ is used to reduce carbon dioxide to carbohydrate; ATP is required in this stage.

1

From the direct products of photosynthesis, green plants can synthesise proteins, polysaccharides, lipids and nucleic acids, all of which contribute to the structure and functioning of cells and organelles. In addition, the sugars and the polysaccharide starch form energy stores. During respiration, the sugars are oxidised to carbon dioxide and water, releasing energy which is stored in molecules of ATP ready for use in cellular activities such as the synthesis of new protoplasm (growth), active transport and movement of protoplasm. Figure 1.1 summarises these activities.

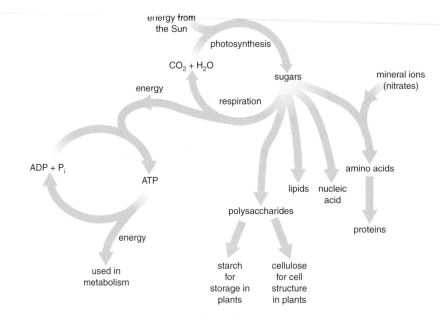

Figure 1.1 Diagram to show how the products of photosynthesis may be used

QUESTION

List the requirements of photosynthesis and their sources.

In addition to the raw materials carbon dioxide and water, a supply of mineral ions is necessary for amino acid and, subsequently, protein formation. Carbon dioxide is obtained from the atmosphere and enters the plants through the stomata on the aerial parts, especially the leaves. The water and mineral ions are obtained from the soil and are transported through the plant in the xylem tissue from the roots to the leaves. Water is necessary for all living processes and there are no special ways in which plants can ensure a supply of water specifically for photosynthesis. Water is continually taken up by green plants to maintain the turgidity of the tissues and to replace that lost in transpiration. The amount of water needed in photosynthesis is small compared with that taken up and lost through transpiration.

Net photosynthesis

In order for growth to occur in a plant, there must be a net gain of energy, so photosynthesis must exceed respiration. The total amount of carbon dioxide 'fixed' as carbohydrate is referred to as **gross photosynthesis**. A certain amount of carbon dioxide will be released as a result of **respiration**. So growth will depend on the difference between gross photosynthesis and respiration, which is known as **net photosynthesis**. More detail about the role of photosynthesis in **productivity** can be found in *Exchange and Transport Energy and Ecosystems*, Chapter 7.

Leaf structure

The main photosynthetic organ of a green plant is the **leaf** (Figure 1.2), although the cortex of green herbaceous stems may contain cells with large numbers of chloroplasts, enabling photosynthesis to occur. In most plants, the leaf provides a large, thin, flat surface, which traps the light effectively. Leaves are particularly well adapted as photosynthetic organs because:

- they provide a large surface area over which light can be absorbed
- they are thin, so the diffusion paths of the gases during gaseous exchange are short: carbon dioxide diffuses in and oxygen diffuses out during daylight
- the **midrib** and extensive network of **veins** provide good support for the thin **lamina**, or **blade**
- the extensive network of veins enables efficient transport of materials to and from the photosynthesising cells; none of the photosynthesising cells is far away from the **xylem** which transports water and mineral ions needed in the process, or from the **phloem** which transports the products of photosynthesis away
- the **mesophyll** tissue, made up of **palisade cells** and **spongy cells**, contains large numbers of **chloroplasts**; the chloroplasts contain the green pigment **chlorophyll**, which absorbs the light energy.

An examination of its internal structure reveals how the leaf is further adapted to the function of photosynthesis. The organisation of the different tissues can be seen in the transverse section of a dicotyledonous leaf in Figure 1.3 and the functions of these tissues are summarised in Table 1.1.

The palisade tissue of the leaf has the greatest concentration of chloroplasts in its cells, so it is here that most photosynthesis takes place. The structure of a palisade cell has been described in *Molecules and Cells*, but it is relevant here to consider how it is adapted to its function. Because of their columnar shape and orientation, the palisade cells appear to be tightly packed in a layer just below the upper epidermis. Careful observation shows that substantial parts of the long sides of these cells are exposed to the intercellular air spaces, so enabling

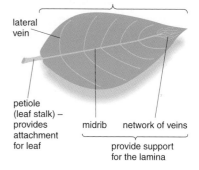

lamina (leaf blade) – large, broad, flat; large surface area for efficient light absorption and gaseous exchange

lateral vein

petiole (leaf stalk) – provides attachment for leaf

midrib   network of veins

provide support for the lamina

(a) **Dicotyledonous leaf** (e.g. privet). Definite upper and lower surface with most stomata on the lower surface

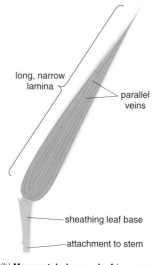

long, narrow lamina

parallel veins

sheathing leaf base

attachment to stem

(b) **Monocotyledonous leaf** (e.g. grasses). No distinct difference between upper and lower surfaces with equal distribution of stomata

Figure 1.2 External features of leaves: (a) dicotyledonous; (b) monocotyledonous

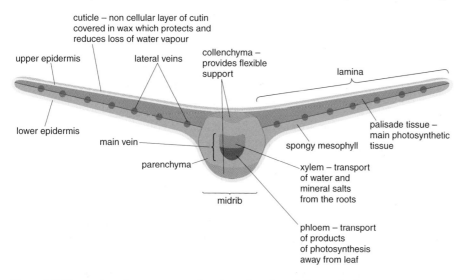

cuticle – non cellular layer of cutin covered in wax which protects and reduces loss of water vapour

upper epidermis   lateral veins

collenchyma – provides flexible support

lamina

lower epidermis

main vein

parenchyma

spongy mesophyll

palisade tissue – main photosynthetic tissue

xylem – transport of water and mineral salts from the roots

midrib

phloem – transport of products of photosynthesis away from leaf

Figure 1.3 Tissue plan of transverse section through a dicotyledonous leaf

PHOTOSYNTHESIS

Table 1.1 *Leaf tissues and their functions*

Tissue	Structure	Function
upper epidermis	layer one cell thick, covered by cuticle; cells flattened; no chloroplasts; usually no stomata present, but if present, fewer than on lower epidermis	cuticle protects and reduces evaporation of water from epidermal cells; transparent so light allowed through to palisade mesophyll
palisade mesophyll	densely packed layer of column-shaped cells; thin cell walls; large numbers of chloroplasts	most photosynthesis occurs in this layer; shape of cells and dense packing means high concentration of chloroplasts; turgidity contributes to support of leaf
spongy mesophyll	irregular-shaped cells containing fewer chloroplasts; form loose network with large air spaces	important for the diffusion of gases; air spaces connect up with stomata; some photosynthesis; turgidity contributes to support of leaf
vascular tissue in midrib and veins	xylem: lignified; vessels and tracheids	xylem conducts water and mineral ions to photosynthesising cells; provides support for the lamina
	phloem: sieve tubes and companion cells	phloem removes the products of photosynthesis (in the form of sucrose)
collenchyma (often found in midrib region)	living cells with extra cellulose thickening in the cell walls	provides flexible support in the leaf, particularly in the midrib region; turgidity also contributes to support
sclerenchyma (often found in midrib region)	lignified cells, fibres; no living contents	provides support in the midrib region
lower epidermis	layer one cell thick; covered by cuticle; cells similar to upper epidermis; many stomata present	important for the presence of stomata allowing gaseous exchange

rapid uptake of available carbon dioxide (Figure 1.4). As the upper epidermis is transparent, the palisade cells are in a position to receive the maximum amount of available light. The cell walls are thin and there is only a thin layer of cytoplasm, allowing rapid diffusion of materials into the chloroplasts.

Chloroplasts and chloroplast pigments

Chloroplasts are saucer-shaped organelles, usually between 4 and 10 μm in diameter and 1 μm thick, surrounded by a double membrane forming the **chloroplast envelope**. Inside the chloroplast is a complex system of internal membranes called **lamellae**, or **thylakoids**, which develop from extensions of the inner membrane of the envelope. They are embedded in a colourless matrix called the **stroma**. In some regions, the thylakoids are arranged in stacks, forming **grana**, which are connected to each other by **stroma thylakoids**. A simplified model of the ultrastructure of a chloroplast is illustrated in Figure 1.5. The membranes which make up the thylakoids, in which the chlorophyll molecules are embedded, consist of approximately equal proportions of protein and lipid. Conversion of light energy to chemical energy (light-dependent reaction) takes place in the thylakoids and the reactions of the light-independent reaction, involving the reduction of carbon dioxide to carbohydrates, take place in the stroma.

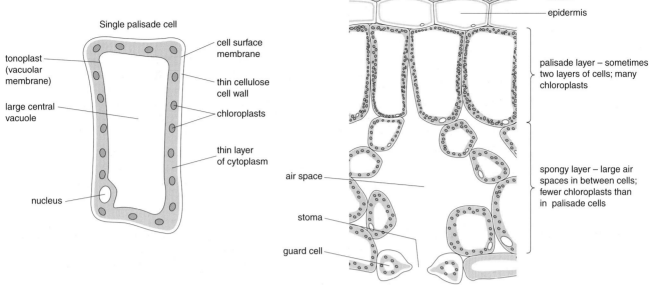

Figure 1.4 Diagram showing cellular structure of leaf tissue with detail of a palisade cell (magnification c. × 500)

The chloroplasts contain small 70S ribosomes and circular DNA, and are similar in structure to some photosynthetic prokaryotes. The presence of these structures has led to the suggestion that the chloroplasts of green plants are the descendants of cyanobacteria, which became incorporated into eukaryotic cells at an early stage in the evolution of green plants. This theory is known as the **endosymbiont theory** and applies to mitochondria as well as chloroplasts.

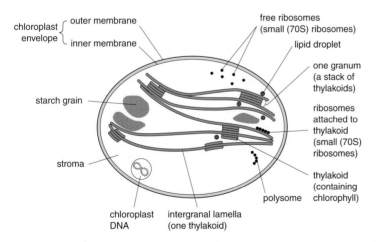

Figure 1.5 Chloroplast ultrastructure (simplified; magnification × 15 000)

All photosynthetic organisms possess at least one pigment which can absorb light and start the reactions of photosynthesis. These pigments can be extracted using organic solvents and separated from each other by chromatography. Three different groups of pigments are found in photosynthetic organisms:

- **chlorophylls**
- **carotenoids**
- **phycobilins**.

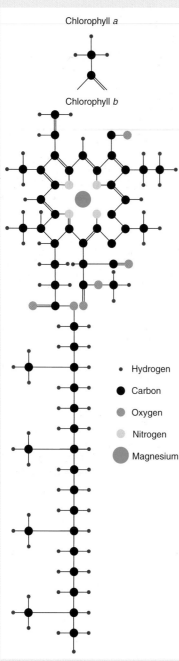

Chlorophyll *a*

Chlorophyll *b*

• Hydrogen
● Carbon
● Oxygen
● Nitrogen
● Magnesium

Figure 1.6 Atomic structure of chlorophyll molecules a *and* b

EXTENSION MATERIAL

Structure of chlorophyll molecules
The molecules have a porphyrin 'head' attached to a long hydrocarbon (phytol) 'tail'. The porphyrin head is polar and consists of a tetrapyrrole ring with a magnesium ion (Mg^{2+}) in its centre, whereas the hydrocarbon tail is non-polar. The porphyrin part of the molecule is bound to the protein of the chloroplast lamellae and the hydrocarbon chain extends into the lipid layer. The chemical structure is shown, for interest, in Figure 1.6.

The **chlorophylls** give leaves their green colour as they absorb red and blue-violet light, reflecting green light. There are several forms of chlorophyll, differing slightly in colour, chemical structure and absorption peaks. The most abundant is **chlorophyll *a***, which plays a central role in the absorption of light energy and is present in all photosynthetic organisms where oxygen is evolved. **Chlorophyll *b*** is also found in the leaves of green plants and in the green algae. Other forms of chlorophyll are characteristic of some groups of algae.

Present in all photosynthetic organisms, the **carotenoids** are yellow or orange pigments which absorb blue-violet light. They are usually masked by the chlorophylls, but their colour shows up in the autumn, when the chlorophylls break down. They are situated quite close to the chlorophyll molecules in the chloroplast lamellae. The light energy which they absorb is transferred to chlorophyll *a*. The most widespread of the carotenoids is **β-carotene**, present in all green plants and algae.

Table 1.2 summarises the various photosynthetic pigments, their occurrence and the range of their light absorption.

ADDITIONAL MATERIAL

Photosynthetic pigments in marine algae (seaweeds)
Fucoxanthin is a yellow-brown pigment, related to the carotenoids, which gives the characteristic colour to brown algae. The **phycobilins**, found in cyanobacteria and red marine algae, are structurally similar to chlorophyll *a*, but do not contain a magnesium atom. They absorb light in the middle of the visible spectrum, which enables the red marine algae to photosynthesise in dim light conditions under water.

Absorption and action spectra
An **absorption spectrum** is a graph of the relative absorbance of different wavelengths of light by a photosynthetic pigment. It is obtained by subjecting

Table 1.2 *Characteristics of photosynthetic pigments*

Pigment	Colour	Occurrence	Wavelengths of light absorbed
chlorophyll *a*	blue-green	green plants, algae	peaks at 420 nm and 660 nm
chlorophyll *b*	yellowish-green	green plants, green algae	peaks at 435 nm and 643 nm
β-carotene	orange-yellow	green plants, algae	between 425 and 480 nm
fucoxanthin	yellow-brown	brown algae (seaweeds)	between 425 and 475 nm
phycobilins	red	red algae, cyanobacteria	between 490 and 576 nm

solutions of each pigment to different wavelengths of light and measuring how much light is absorbed.

As shown in Figure 1.7, chlorophyll *a* absorbs blue-violet light (420 nm) and red light (660 nm). Very little absorbance occurs between 500 and 600 nm, so light of these wavelengths will be reflected. The absorption spectrum for chlorophyll *b* is very similar.

An **action spectrum** is a graph showing the amount of photosynthesis at different wavelengths of light. It can be obtained by allowing plants, such as Canadian pond weed, to photosynthesise for a stated time in light of each wavelength in turn and measuring the volume of gas evolved. A graph is then plotted of rate of photosynthesis against wavelength of light.

If an action spectrum and the absorption spectrum for an extract of chlorophyll pigments are plotted on the same graph, as shown in Figure 1.8, it can be seen that they resemble each other closely, indicating that there is good evidence for these pigments being involved in the absorption of light for photosynthesis.

Light-dependent reaction

In this stage, energy is transferred from light into ATP and reduced coenzyme, $NADPH+H^+$, is formed. In order to appreciate how these events take place, it is necessary to understand how the chlorophyll molecules are arranged and their association with the thylakoid membranes.

The photosynthetic pigment molecules are organised into two photosystems: **photosystem I** (**PSI**) and **photosystem II** (**PSII**). These photosystems can be seen on the thylakoid membranes as different-sized particles, which were first discovered when chloroplasts were subjected to a technique known as 'freeze-fracturing' before being viewed using electron microscopy. It appeared that the larger particles, now known to be PSII, were associated with the granal thylakoids, and the smaller particles, PSI, with the stromal thylakoids (Figure 1.9). Each photosystem consists of a large number of **accessory pigment** molecules, such as chlorophyll *b* and the carotenoids, which absorb light energy and transfer it to a **reaction centre** consisting of one of the **primary pigment** molecules, which are specialised forms of chlorophyll *a*.

EXTENSION MATERIAL

In PSI, the primary pigment molecule is P700 and in PSII, the primary pigment molecule is P680. The accessory pigments include all the other forms of chlorophyll (other forms of chlorophyll *a* with different absorption peaks and chlorophyll *b*) and the carotenoids. They pass on their energy to the primary pigments. The primary pigments then emit the electrons which cause the light-dependent reactions of photosynthesis.

Chlorophyll molecules contain electrons that can be excited by light energy to become **high energy electrons**. When light is absorbed, these electrons are emitted and are taken up by other molecules, known as **electron acceptors** or **electron carriers**, which pass them on to another molecule. The chlorophyll molecule, having lost electrons, is left in an oxidised state and

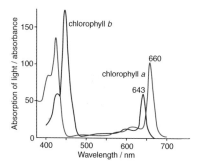

Figure 1.7 Absorption spectra for chlorophylls a *and* b

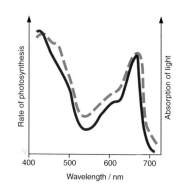

Figure 1.8 Graph comparing absorption spectrum (solid line) and action spectrum (broken line)

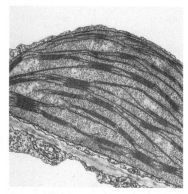

Figure 1.9 Electronmicrograph of chloroplast showing granal and stromal thylakoids

PHOTOSYNTHESIS

the electron carrier is reduced. A word equation for the effect of light on chlorophyll can be written as follows:

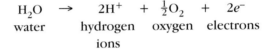

$$\underset{\text{(reduced)}}{\text{chlorophyll}} \xrightarrow{\text{light energy}} \underset{\text{(oxidised)}}{\text{chlorophyll}^+} + \underset{\text{(electron)}}{e^-}$$

The electrons are passed along a chain of electron carriers in a series of **oxidation–reduction** (**redox**) reactions. Each carrier in the series is at a lower energy level than the one preceding it and, as the electrons are passed along, sufficient energy is released to build up molecules of ATP from ADP and inorganic phosphate. As this process uses light energy, it is referred to as **photophosphorylation**, to distinguish it from **oxidative phosphorylation** which occurs as a result of respiration.

In **non-cyclic photophosphorylation**, light energy is absorbed by the chlorophyll molecules in PSI and PSII. High energy electrons are emitted from the chlorophyll pigments in both photosystems, leaving them both oxidised. In order for more light energy to be absorbed, the electrons need to be replaced. Associated with PSII is an enzyme which catalyses the splitting of water into hydrogen ions, electrons and oxygen:

$$\underset{\text{water}}{H_2O} \to \underset{\substack{\text{hydrogen} \\ \text{ions}}}{2H^+} + \underset{\text{oxygen}}{\tfrac{1}{2}O_2} + \underset{\text{electrons}}{2e^-}$$

The electrons released from the water replace the electrons emitted from the chlorophyll molecules in PSII, the hydrogen ions are released into the lumen of the thylakoid and the oxygen is released as a waste product. The two electrons emitted from the chlorophyll molecules in PSII are picked up by electron acceptor A and passed along a chain of carriers. In this process, some of the energy from these electrons is used to move hydrogen ions across the thylakoid membrane from the stroma. Eventually they replace the electrons emitted from the chlorophyll molecules in PSI, thus restoring stability to the chlorophyll molecules in PSI.

The electrons emitted from the chlorophyll molecules in PSI are passed to electron acceptor B and used to reduce NADP in the stroma. This reaction also removes hydrogen ions from the stroma.

The thylakoid membranes are impermeable to hydrogen ions and there is a pH difference between the lumen of the thylakoid and the stroma, with a high concentration of hydrogen ions inside the lumen and a low concentration of hydrogen ions in the stroma. On the thylakoid membranes are complex protein molecules made up of two protein parts, one acting as a pore and the other acting as ATP synthetase (ATPase). As the hydrogen ions move from the lumen of the thylakoid through the channel created by the pore, energy is transferred from the hydrogen ions and used to build up ATP from ADP and inorganic phosphate, catalysed by ATP synthetase. The hydrogen ion concentration gradient across the thylakoid membrane is maintained by:

- the splitting of water
- the transport of electrons from PSII along the chain of carriers
- the formation of $NADPH + H^+$.

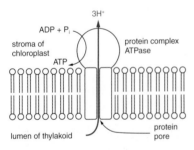

One ATP molecule is built up for every 3 H^+ ions passing through the pore down the gradient

path of hydrogen atoms

Figure 1.10 Diagram of thylakoid membrane to illustrate the role of protein molecules in ATP formation

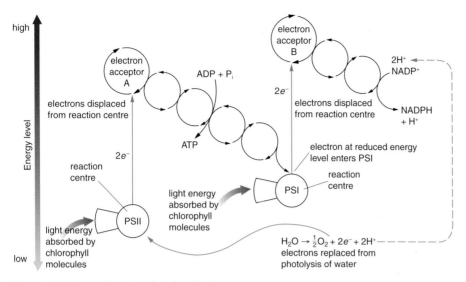

Figure 1.11 Non-cyclic photophosphorylation

The events described above are illustrated by Figure 1.10 and Figure 1.11, and result in the production of NADPH+H$^+$ and ATP, both of which are used in the light-independent stage.

Cyclic photophosphorylation, in which only ATP is formed, involves PSI only. Light absorbed by the chlorophyll molecules in PSI causes electrons to be emitted, which are picked up by a different electron carrier. The energy in the electrons is used to move hydrogen ions across the thylakoid membrane into the lumen. The electrons are then passed back into the chlorophyll molecules in PSI to replace the electrons that were emitted. ATP is synthesised as described above. Figure 1.12 illustrates this.

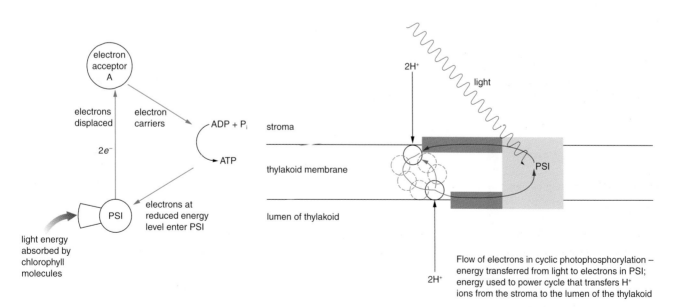

Figure 1.12 Cyclic photophosphorylation

PHOTOSYNTHESIS

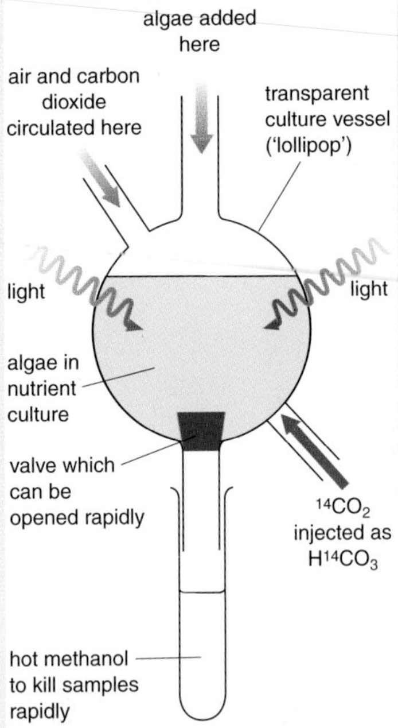

Figure 1.13 Diagram illustrating Calvin's 'lollipop' apparatus to determine the path taken by carbon dioxide in photosynthesis. Carbon dioxide containing radioactive carbon is bubbled through the algal suspension in the flask.

DEFINITION

The Law of limiting factors
When a chemical process, such as photosynthesis, is dependent on a number of factors being favourable, the rate of the reaction is limited by that factor which is nearest its minimum value.

Light-independent reaction

This stage of photosynthesis takes place in the stroma of the chloroplast. **NADPH+H$^+$** from the light-dependent stage is used to reduce **carbon dioxide** to **carbohydrate** and the **ATP** is used to provide the necessary energy. The sequence of enzyme-controlled reactions in this stage was determined by **Calvin** and his associates, using the alga *Chlorella* and the radioactive isotope ^{14}C. The reactions of the light-independent stage are therefore known as the Calvin cycle. His experiment is summarised in Figure 1.13.

Calvin identified the first product of photosynthesis as **glycerate 3-phosphate (phosphoglyceric acid)**, a 3C organic acid, which is reduced to **glyceraldehyde 3-phosphate** from which other carbohydrates can be synthesised.

Carbon dioxide diffuses into the cells of the palisade tissue and into the chloroplasts down a concentration gradient. In the stroma of the chloroplast, a molecule of carbon dioxide combines with a **carbon dioxide acceptor molecule**: a 5C pentose, **ribulose 1,5-bisphosphate (RuBP)**. Two molecules of glycerate 3-phosphate are formed. This is a **carboxylation** (addition of carbon dioxide) and the enzyme responsible for catalysing the reaction is the **carboxylase, ribulose bisphosphate carboxylase (RuBisCO)**. The glycerate 3-phosphate is first phosphorylated, using ATP, then reduced to glyceraldehyde 3-phosphate, using NADPH+H$^+$. This process is summarised in Figure 1.14.

Glyceraldehyde 3-phosphate is used for two purposes:
- to regenerate the carbon dioxide acceptor molecule RuBP so that the fixation of carbon dioxide can continue
- to build up more complex organic compounds, such as sugars, starch and amino acids.

The regeneration of RuBP involves a complex series of reactions in which 3C, 4C, 5C, 6C and 7C sugar phosphates are formed. For every six molecules of glyceraldehyde 3-phosphate formed during the Calvin cycle, only one can be used for the synthesis of product as five are needed to regenerate a molecule of RuBP. Some starch may be built up in the chloroplast, but the synthesis of sucrose takes place in the cytoplasm. The glyceraldehyde 3-phosphate is able to leave the chloroplast via a specific transport protein in the inner chloroplast membrane. It may then be built up into hexose phosphates and eventually sucrose or starch, or it may be used in the synthesis of fatty acids or amino acids. Figure 1.15 summarises the fate of glyceraldehyde 3-phosphate. Recent research has indicated that different end products are made under different conditions. When light intensity and carbon dioxide concentration are high, then sugars and starch are formed, but in low light intensities amino acids are formed.

Environmental factors affecting photosynthesis

Photosynthesis involves a series of reactions and the overall rate at which the process occurs will be dependent on the rate of the slowest of these reactions. If the light intensity is low, then the rate at which NADPH+H$^+$ and ATP are

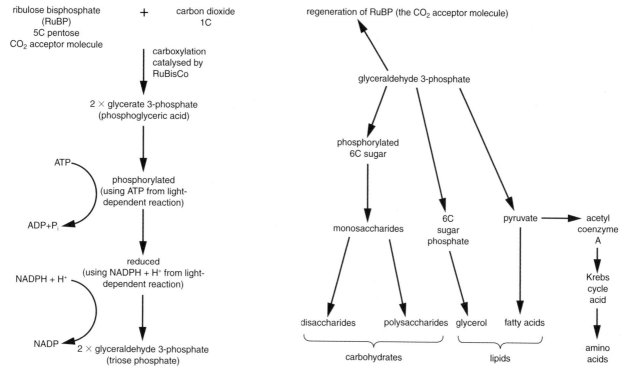

Figure 1.14 Equation summarising the formation of glyceraldehyde 3-phosphate

Figure 1.15 Summarising the fate of glyceraldehyde 3-phosphate

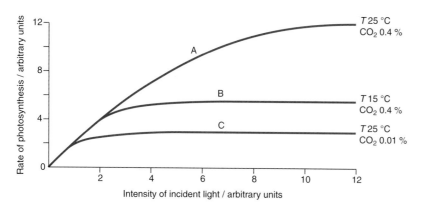

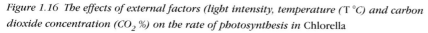

Figure 1.16 The effects of external factors (light intensity, temperature (T °C) and carbon dioxide concentration (CO₂ %) on the rate of photosynthesis in Chlorella

being produced in the light-dependent stage will affect the rate of the reactions in the light-independent stage. In these circumstances, light is said to be a **limiting factor**. If the light intensity is increased, then the rate of photosynthesis will increase until another factor, such as temperature or carbon dioxide concentration, becomes limiting (Figure 1.16).

The chief external factors which affect the rate of photosynthesis are **light intensity, carbon dioxide concentration** and **temperature**.

Light intensity

No photosynthesis occurs in the absence of light, but respiration continues and so the net gas exchange of a green plant will show uptake of oxygen and

QUESTION

The effects of temperature and carbon dioxide concentration on the rate of photosynthesis as the intensity of the incident light increases are shown in Figure 1.16. Using the information on the graph suggest explanations for the differences between the rates shown by the curves A, B and C. Suggest what might happen in A if the light intensity was increased beyond 12 arbitrary units. What would be the effect of increasing the temperature to 35 °C while keeping the carbon dioxide concentration at 0.4 %?

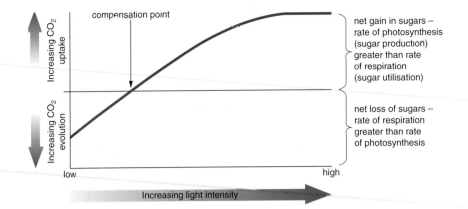

Figure 1.17 Light intensity and the compensation point

release of carbon dioxide. Carbohydrates are being used up in these conditions. At very low light intensities, some photosynthesis will occur using the carbon dioxide released by respiration, so the net gas exchange will still show uptake of oxygen and release of carbon dioxide. As the light intensity increases, so the rate of photosynthesis increases until the amount of carbon dioxide released from respiration is equal to the amount used up in photosynthesis. At this light intensity, known as the **light compensation point** (Figure 1.17), the rate of carbon dioxide production during respiration is equal to the rate at which carbon dioxide is taken up for photosynthesis.

At light intensities higher than this, there will be a net uptake of carbon dioxide and release of oxygen and the amount of carbohydrate in the plant will increase.

As the light intensity increases further, the rate decreases and then reaches a plateau, where further increase in light intensity has no effect on the rate of photosynthesis. At this stage, either another factor has become limiting or light saturation has been reached.

Prolonged exposure to strong sunlight may cause damage to the chloroplast pigments. If this happens, the leaves appear bleached and eventually die.

Carbon dioxide concentration

Carbon dioxide is needed in the light-independent reactions, where it is involved in the formation of carbohydrates. The carbon dioxide concentration of the atmosphere is about 0.035 per cent, or 350 parts per million, by volume, and under normal conditions this is the factor which limits the rate of photosynthesis. Increasing the concentration of carbon dioxide will increase the rate of photosynthesis and so it is possible to increase the growth of glasshouse crops, such as tomatoes and lettuces, by enriching the atmosphere with carbon dioxide.

Temperature

Temperature affects the rate of chemical reactions and is important in the light-independent stage of the photosynthetic process because the reactions are enzyme controlled. In temperate climates, the optimum temperature for photosynthesis is about 25 °C, but an increase of 10 °C will double the rate, provided no other factor becomes limiting. Temperature does not affect the rate of the light-dependent reaction.

Mineral nutrition

In addition to carbon, hydrogen and oxygen, which are obtained from carbon dioxide and water, green plants need at least 13 essential elements in order to produce new tissues and maintain their correct functioning. **Nitrogen** and **sulphur** are essential constituents of amino acids, **phosphorus** is needed for the synthesis of nucleic acids and phospholipids, **magnesium** is a constituent of chlorophyll, and **potassium** is an important constituent of cell sap. These elements are taken up by the roots as inorganic ions from the soil water: nitrogen as nitrate ions (NO_3^-), sulphur as sulphate ions (SO_4^{2-}), phosphorus as phosphate ions (PO_4^{3-}), magnesium as magnesium ions (Mg^{2+}), and potassium as potassium ions (K^+). Some of the functions of nitrates, phosphates and magnesium are shown in Table 1.3.

Most absorption occurs in the root hair region of the roots, situated a few centimetres from the root apex. In this region, the cells in the outermost layer, called the piliferous layer, have long hair-like extensions which penetrate between the mineral particles of the soil. These mineral particles are surrounded by films of water in which the mineral ions are dissolved. The cellulose cell walls of the root hair cells are freely permeable, so water and mineral ions diffuse through and are then taken up into the root hair cells, either by diffusion or by active transport.

The movement of ions across membranes is affected by a combination of the concentration gradient and an electrical gradient known as the **electrochemical gradient**. The ions will be attracted to areas of opposite charge and will move away from areas of similar charge. Where the concentration of an ion in the root cells is less than its concentration in the soil water, it will diffuse into the root down the electrochemical gradient. In many cases, however, the uptake of ions involves movement from a lower concentration in the soil water to a higher concentration in the plant tissues. This is an active process and requires energy from respiration. The relative concentrations of different ions inside cells differs from their relative concentrations in the soil water, providing evidence for selective uptake. It appears that the cells will take up those ions they require in preference to others. The cell membranes are very effective diffusion barriers and the active transport of ions is associated with specific ion pumps located on them.

Once inside the root, the ions may be transported towards the centre of the root in the cytoplasm of the cortex cells, the **symplast pathway.** They pass from cell to cell by means of the **plasmodesmata**, which are fine cytoplasmic connections between neighbouring cells. When the xylem is reached, it is thought that the ions are actively secreted into this tissue, from where they are transported, in the **transpiration stream**, to the aerial parts of the plant. Alternatively, the ions may be transported by the **apoplast pathway**, through the cell walls, partly by diffusion and partly with the flow of water due to transpiration. Water and ion uptake by the roots of plants is described in more detail in *Exchange and Transport, Energy and Ecosystems*.

Table 1.3 *Functions of nitrate, phosphate and magnesium ions*

Mineral ion	Functions
nitrate [NO_3^-]	essential for the synthesis of amino acids, proteins, nucleic acids, pigment molecules, coenzymes
phosphate [PO_4^{3-}]	required for synthesis of nucleic acids, phospholipids; component of nucleotides (ATP); phosphate group involved in energy transduction; phosphorylated intermediates in metabolism
magnesium [Mg^{2+}]	constituent of chlorophyll molecule; required as an activator for many dehydrogenase and phosphate transfer enzymes

PHOTOSYNTHESIS

Chromatography of chloroplast pigments

Introduction

Chloroplasts contain a number of different pigments, including chlorophyll *a*, chlorophyll *b*, β-carotene and xanthophylls such as lutein. These can be extracted using organic solvents and separated using the technique of chromatography. Thin-layer chromatography, using small silica gel plates, gives excellent and rapid separation. *CAUTION: the solvents used in this practical are highly flammable and harmful. Use of a fume cupboard is recommended.*

Materials

- Plastic-backed silica gel chromatography plates, cut to the size of microscope slides
- Fresh leaf material, such as spinach
- Mortar and pestle
- Fine sand
- Scissors
- Stoppered test tube
- Propanone
- Distilled water
- Petroleum ether (40 to 60 °C fraction)
- Pipette
- Small sample bottle
- Fine glass tube, such as a melting point tube
- Glass beaker covered with suitable lid
- Solvent: hexane-diethyl ether-propanone in the proportions 60 : 30 : 20 by volume (see **Note** in the margin)

Method

1. Use scissors to cut up about 1 g of fresh leaf tissue and grind thoroughly in a mortar with some fine sand.
2. Transfer the ground-up material to a stoppered test tube, add 4 cm³ of propanone, shake thoroughly then leave to stand for 10 minutes.
3. Add 3 cm³ of distilled water and shake.
4. Now add 3 cm³ of petroleum ether, shake gently, then allow the solvents to separate. The pigments will collect in the top layer of petroleum ether, which can be removed with a pipette and transferred to a small sample bottle.
5. Prepare a silica gel chromatography plate by carefully drawing a pencil line origin about 1.5 cm from the bottom. Avoid touching the surface of the plate with your fingers.
6. Use a fine glass tube to transfer the extract to the origin. Briefly touch the surface of the plate with the end of the tube, allow the petroleum ether to

evaporate and repeat this process several times to produce a small, concentrated spot of pigment (Figure 1.18).
7. Stand the plate in a small, covered beaker containing solvent (hexane-diethyl ether-propanone). The level of solvent **must** be below the level of the pigment spot. Allow the solvent to rise until the solvent front reaches the top 1 to 2 cm of the plate. Do not allow the solvent to reach the top edge.
8. Remove your completed chromatogram and allow the solvent to evaporate. Make a careful drawing to show the positions and colours of the separated pigment spots.

Results and discussion

The separated pigments may be identified by their **Rf values**. The Rf value is defined as the distance moved by the spot (measured from the origin to the centre of the spot) divided by the distance moved by the solvent (measured from the origin to the solvent front). Rf values for some chloroplast pigments are given in Table 1.4.

1. Were all the leaf pigments removed by this technique?
2. How many pigments were you able to identify in your extract?

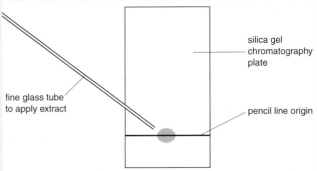

Figure 1.18 Extracting pigments from chloroplasts

silica gel chromatography plate

pencil line origin

fine glass tube to apply extract

Table 1.4 *Rf values of some chloroplast pigments*

Pigments	Rf values
β-carotene	0.98
phaeophytin	0.53
chlorophyll *a*	0.42
chlorophyll *b*	0.34
xanthophylls	0.30 – 0.07

HIGHLY FLAMMABLE
propanone
petroleum ether
hexane

EXTREMELY FLAMMABLE
diethyl ether

HARMFUL
diethyl ether
(intoxicating vapour)
hexane
petroleum ether

Note: If no fume cupboard is available it is recommended that propanone (90%) – petroleum ether (80 to 100°C fraction) in the proportions 1:9 by volume is used as the solvent

PRACTICAL | **The effects of light intensity and carbon dioxide concentration on the rate of photosynthesis**

Introduction

A convenient way of measuring the rate of photosynthesis is to find the rate of production of oxygen bubbles from a submerged aquatic plant, such as *Elodea canadensis* (Canadian pond weed). A freshly cut shoot of *Elodea*, supported by tying carefully to a glass rod using cotton, can be inverted in a large beaker full of water. The cut end should be no more than about 1 cm below the surface of the water. When illuminated, a stream of bubbles should emerge from the cut end of the shoot. A tiny amount of detergent added to the water will allow the bubbles to escape freely. These bubbles are not pure oxygen (nitrogen and carbon dioxide are also present), but the rate of bubbling will give a good indication of the rate of photosynthesis.

A more quantitative method involves collecting the bubbles produced in a standard time and measuring their total volume. The apparatus shown in Figure 1.19 (known as a photosynthometer) can be used for such quantitative studies.

Materials

- Photosynthometer
- 60 or 100 W bench lamp. (If available, a slide projector with a heat shield provides a high intensity light source)
- Fresh *Elodea* (if possible, this should be kept in dilute sodium hydrogencarbonate solution and brightly illuminated before the experiment)
- Stop clock
- Metre rule
- Sodium hydrogencarbonate

Method

1 Remove the plunger from the syringe and completely fill the apparatus with water from a tap. Replace the plunger.
2 Assemble the apparatus as shown in Figure 1.19. Push in the plunger almost to the end of the syringe and ensure that there are no air bubbles in the tubing.
3 Add a pinch of sodium hydrogencarbonate to the test tube containing the *Elodea* and stir to dissolve. This will provide carbon dioxide and ensure that it is not a limiting factor.
4 When ready to start the experiment, darken the laboratory and start with the bench lamp close to the *Elodea*. Allow time for the *Elodea* to equilibrate, then collect the oxygen given off for a suitable period of time.

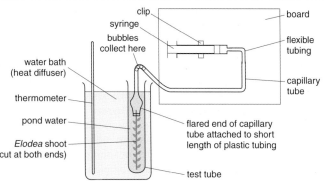

Figure 1.19 A photosynthometer

5 Using the plunger, carefully draw the bubbles into the capillary tube so that the volume or length of the bubble can be measured.
6 Take two more readings at this distance.
7 Repeat this procedure with the lamp at different, measured, distances from the *Elodea*, such as 10, 15, 20, 30, 40 and 80 cm. In each case allow time for the *Elodea* to equilibrate.
8 Record the distances and corresponding volume of oxygen produced. Check the thermometer to ensure that the temperature of the water remains constant (why?).

Results and discussion

1 The light intensity is inversely proportional to the square of the distance (d) between the light source and the *Elodea*. You can calculate the relative light intensity by $1/d^2$. Since the figures for light intensity are small, each one can be multiplied by 10 000. This makes them easier to plot.
2 Make a table to show the mean rate of photosynthesis (volume of oxygen produced per unit time) at each light intensity.
3 Plot a graph to show the relationship between the rate of photosynthesis and light intensity. Comment on the shape of your graph.

Further work

The same apparatus can be used to investigate the effect of other factors, such as carbon dioxide concentration or temperature, on the rate of photosynthesis. Carbon dioxide concentration can be varied by using a range of sodium hydrogencarbonate solutions, for example, 0.1, 0.2, 0.3, 0.4 and 0.5 %. Careful experimental design is necessary to keep other factors constant.

PRACTICAL | Investigating the effects of mineral deficiency

Introduction

The aim of this practical is to investigate the effects of mineral deficiency on the growth of plants. Seedlings are grown in solutions containing a range of mineral salts, including those which lack phosphate, nitrate, calcium, potassium, magnesium, iron and sulphate, plus the complete medium.

Suitable seedlings which can be used for this experiment include: maize (*Zea mays*), castor beans (*Ricinus communis*), tomato (*Lycopersicon esculentum*) and cabbage (*Brassica oleracea*). Seeds are germinated in moist vermiculite, then transferred to the nutrient solution, as shown in Figure 1.20. At weekly intervals, the plants are measured and any deficiency symptoms noted.

Materials

- Mineral-deficiency water culture media (available from Philip Harris Education)
- Seedlings germinated in moist vermiculite and grown to the first leaf stage
- Very clean test tubes or boiling tubes, and rack
- Aluminium foil
- De-ionised water

Method

1 Make up the culture media and set up eight tubes containing the following:
- complete (normal) medium
- medium lacking phosphate
- medium lacking nitrate
- medium lacking calcium
- medium lacking potassium
- medium lacking magnesium
- medium lacking iron
- medium lacking sulphate.
2 Wrap each tube in aluminium foil.
3 Select eight seedlings and set up the cultures as shown in Figure 1.20.
4 Leave the experiment at room temperature and each week record the following:
- shoot and root lengths
- leaf number and size
- internode number and length
- deficiency symptoms, such as changes in shape or yellowing of leaves.

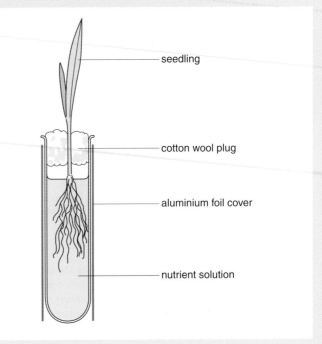

Figure 1.20 Water culture experiment to investigate the effects of mineral deficiency

Results and discussion

1 Record all your results in a table.
2 Present your results as graphs to show growth patterns.
3 Discuss your results as fully as you can.
4 Explain why the tubes were wrapped in aluminium foil.
5 Make a table to show the functions of the ions in the culture media.

Further work

1 Find out the composition of plant mineral culture media, including macronutrients and micronutrients.
2 Devise a key to mineral ion deficiency symptoms.
3 You could investigate the effects of mineral deficiency on the growth and development of rapid cycling brassicas.
4 Find out about the use of **hydroponics** for the growth of glasshouse crops.

Control of growth in plants

Plant growth and development may be influenced by many internal and environmental factors. In this chapter, the influence of light as one of the major environmental factors and the effects of a range of plant growth substances (internal factors) are considered.

- **Light** is an essential environmental factor in the growth and development of plants. It can be detected by **phytochrome**, a **photoreceptor** pigment, and it affects aspects of development such as germination, the initiation of flowering and the direction of plant growth.
- **Internal control of growth and development** is achieved by chemical substances synthesised in the plant. These substances are known as **plant growth substances** and have certain features in common with animal hormones. They may be produced in one part of the plant and translocated to another part, where they create a response. They normally act at low concentrations and their actions may promote or inhibit certain growth processes. Much interest lies in the production of artificial growth substances (often called **growth regulators**) and in their commercial applications.

Detection of light in flowering plants

The detection of light is known as **photoreception**. It involves the absorption of light by a pigment, known as a **photoreceptor**. Many aspects of plant growth and development are known to be controlled by light.

Plants grown entirely in the dark have a distinctive appearance. They
- appear yellowish-white in colour because chlorophyll is not formed and no chloroplasts develop
- have more elongated stems than normal
- have tiny, unexpanded leaves
- have hooked tips to the stems
- are fragile and die as soon as all the food reserves in the seed are used up.

Such plants are described as **etiolated** (Figure 2.1) and their appearance indicates that exposure to light is necessary for the development of chlorophyll, for the expansion of the leaves and for the upright growth of the stem tips. Without chlorophyll and light, the plant cannot photosynthesise and so, when all the food reserves have been exhausted, death is inevitable.

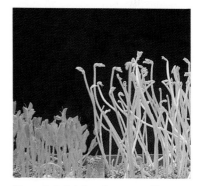

Figure 2.1 Etiolated cress seedlings (right) alongside normal plants (left)

Certain varieties of seeds require the stimulus of light to initiate germination. This topic was discussed in detail in *Respiration and Coordination* Chapter 2, to which reference should be made.

The experimental work on the initiation of seed germination led to the discovery of the photoreceptor pigment, **phytochrome**. This pigment was isolated and identified by Borthwick and Hendrick. It is a blue-green pigment

CONTROL OF GROWTH IN PLANTS

QUESTION

Design an experiment which you could carry out to determine how long seeds need to be exposed to the correct wavelength of light before germination is initiated.

present in the leaves of plants in very small quantities. It exists in two interconvertible forms, P_R or P_{660}, absorbing red light of wavelength 660 nm, and P_{FR} or P_{730}, absorbing far-red light of wavelength 730 nm.

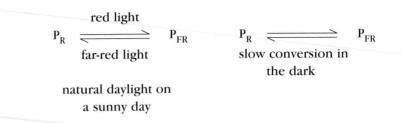

Some plants need to be exposed to the correct photoperiod (day length) before flower buds are initiated. In **long-day plants**, such as wheat, red clover and spinach, flowering is initiated by exposure to cycles of long days and short nights, whereas **short-day plants**, such as chrysanthemum, morning glory and tobacco, flower after exposure to cycles of short days and long nights. Some plants, such as tomato, antirrhinum and balsam, do not appear to be affected by the day length and are described as **day-neutral**. These plants usually flower when they reach a certain size.

It has been suggested that phytochrome is involved in the photoperiodic response to flowering. P_{FR} is known to be the form of phytochrome that promotes flowering in long-day plants, but it inhibits flowering in short-day plants. The exact mechanism is not understood, but the levels of P_{FR} could affect other compounds, such as plant growth substances, which are also known to have an effect. What is clear is that exposure of the leaves to the correct photoperiod triggers the response.

There is evidence to suggest that phytochrome is involved in the unbending of the plumule hook in the stems of seedlings of French beans after germination and also in the initial expansion of leaves. These responses are similar to germination in that they appear to be triggered by a high level of P_{FR}.

Only a short exposure to light of the right wavelength is necessary to trigger germination in lettuce seeds, but much longer periods of illumination are required for leaves to expand fully and for the synthesis of chlorophyll. From a number of investigations, it has been shown that red light is more effective at inducing rapid responses, whereas far-red light is more effective in stimulating long-term responses. The explanation for this apparently paradoxical state may be that long-term responses require high levels of P_R or low levels of P_{FR}, or possibly that the level of P_{FR} has to be maintained within certain limits.

The effect of light on the growth of plants

Plants respond to external stimuli, such as light, by means of growth movements called tropisms. The bending of seedlings towards a light source is an example of a tropism and is achieved by unequal growth of the two opposite sides of the stem of the seedling. Tropic growth movements are directed by an external stimulus. They are described as positive if growth is towards the stimulus or negative if growth is away from the stimulus. Thus shoots are **positively phototropic**, i.e. they grow towards light.

Phototropism, the response of plant organs to the direction of light, was first investigated by Charles Darwin and his son Francis in 1880, using oat **coleoptiles** (Figure 2.2). Much of the experimental work has been carried out using these first shoots of seeds of the grass family. The coleoptile is the sheath covering the plumule, or first shoot, of the germinating embryo. Inside the coleoptile lie the **apical meristem** and **leaf primordia**. These parts contain actively dividing cells, and these are the cells which differentiate to become new structures in the shoot. The leaf primordia contain the cells which develop into leaves. Coleoptiles are useful for experimental work because they are very sensitive to light and, in the dark, they grow at first like a stem. Oat coleoptiles were convenient to use because they could be grown easily in large numbers and they showed the same responses as shoots.

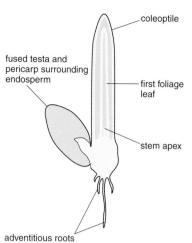

Figure 2.2 Oat seedling showing coleoptile

BACKGROUND

The Darwins observed that, when oat coleoptiles were grown in conditions where they were illuminated from one side, the coleoptiles bent towards the light. The curvature occurred in the region of growth (elongation) just below the tip. They then removed the tips of the coleoptiles and showed that the response did not occur. In further experiments they covered the tips with opaque caps and also buried the coleoptiles, leaving only the tips exposed. In the first case, no curvature occurred, but in the second, it did. Darwin's investigations are summarised in Figure 2.3.

It was suggested that the tips of the coleoptiles were the 'sites of perception' and that some 'influence' passed from the tip to the region of growth, the 'effector'. In 1913, the experiments carried out by a Danish physiologist, Boysen-Jensen, provided further evidence that some kind of signal passes from the tip, where the stimulus is received, to the region of elongation where the response occurs. From his experiments, it appeared that the signal was likely to be a chemical substance. The signal was unable to pass through the impermeable mica plates he inserted in the coleoptiles. The signal also appeared to be produced on the shaded side of the coleoptile, suggesting that light affected its production or distribution (Figure 2.4).

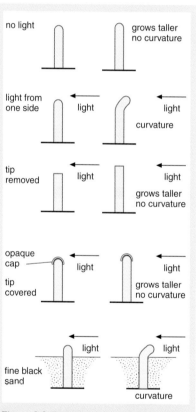

Figure 2.3 Darwin's experiments with oat coleoptiles

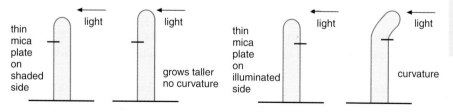

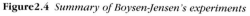

Figure 2.4 *Summary of Boysen-Jensen's experiments*

The existence of a chemical transmitter was proved in 1928 by Fritz Went, a Dutch physiologist. He cut the tips off oat coleoptiles and left them on agar for several hours, so that any substance produced in the tips could diffuse into the agar. Went also showed that exposing coleoptile tips to unilateral light resulted in more of the chemical transmitter being present on the shaded side than on the illuminated side.

The chemical transmitter was named **auxin** and later identified as **indoleacetic acid** (**IAA**). At the time, it was suggested that unilateral light causes auxin, produced at the shoot apex, to move to the shaded side, where the greater concentration stimulates cell elongation in the region just behind the apex, the zone of elongation. This causes the tip of the shoot to bend towards the light.

Plant growth substances

Generally, five groups of plant growth substances are recognised:
  • **auxins**, **gibberellins**, **cytokinins**, **abscisic acid** and **ethene**.

QUESTION

We can see parallels with animal hormones. As you go through this section, look for ways that plant growth substances are similar to animal hormones and ways in which plant growth substances differ from animal hormones.

Auxins, gibberellins and cytokinins act as growth **promoters**, involved in the processes of cell division, cell elongation, differentiation of cells and the initiation of organs. Ethene and abscisic acid act mainly as growth **inhibitors**, associated with processes such as ripening and leaf fall occurring later in plant development. In addition to their individual effects, the plant growth substances can interact with each other, either **synergistically** (whereby the effects are enhanced by two or more growth substances acting together) or **antagonistically** (whereby the effects counteract each other).

Experimental work with auxins and other plant growth substances often relies on exogenous (external) application of the substance, say, on sections taken from stems or coleoptiles. Care must be taken in the interpretation of these experiments in that they provide only indirect evidence for the activity of the growth substance within the living material. This is partly because of the very low concentrations at which growth substances occur and is also a reflection of the complex interactions of one plant growth substance with others.

Auxins

The term 'auxin' now refers to a group of plant growth substances involved generally with elongation growth in coleoptiles and many stems. Auxins also affect other processes, such as initiation of roots, dormancy in lateral buds and fruit development. There are a number of naturally occurring auxins, of which **indoleacetic acid** is the most important. Some synthetic auxins have important applications on a commercial scale in horticulture and agriculture.

IAA is synthesised in regions of active cell division and enlargement, notably the apical meristems (at the tip) of shoots and in young leaves. Transport of IAA away from the apex to older organs is in one direction (polar) and is by an active process which requires energy, with the auxin passing from cell to cell. Transport from sites of synthesis in leaves appears to be non-polar, and occurs in the phloem. This is primarily a passive process, allowing the auxin to move up or down the plant at speeds considerably faster than that achieved by the polar transport.

Auxins promote growth by increasing the plasticity of the cell wall. As the cell wall softens, the cell becomes less turgid and so takes up more water, resulting in expansion of the cell. In the apical growing regions, because of the orientation of cellulose microfibrils in the cell wall, cell expansion tends to be in a longitudinal direction, leading to elongation. There is evidence to suggest that the softening of the cell wall is linked to an **acidification** process in which hydrogen ions (H^+) are pumped, by an active process, out of the cell. The resulting decrease in pH activates enzymes which promote breakdown of bonds within the cell wall structure, leading to the loosening of the cell wall. The process of wall expansion is irreversible and auxin may also promote the synthesis of new wall material, ensuring continued growth. It is likely that, in addition to this rapid short-term effect on cell elongation, auxin also exerts an effect through the activation or repression of specific genes involved in growth responses.

The experiments described earlier in this chapter show how auxins are linked with the phototropic response. Shoots are positively phototropic and the curvature towards the light is the result of increased elongation of the cells in the growing region on the shaded side. This is brought about by a redistribution of auxin laterally across the shoot. Similarly, the upward curvature of horizontal shoots in response to gravity is a negatively geotropic one and is consistent with an accumulation on the lower side of the shoot. Roots are positively geotropic and the downward curvature of horizontal roots is due to a net increase in growth on the upper side of the root. This conflicts with the behaviour of auxins in shoots. The explanation may lie in the concentration of auxin being less than that required for growth in the root cells or to other plant growth substances being involved. Abscisic acid is known to have an inhibitory effect on growth and its accumulation in the region could reduce elongation of the lower surface.

Other activities of auxins are summarised below, together with some examples of applications in horticulture and agriculture.

- **Apical dominance** – auxin in the apical bud has an inhibitory effect on the growth of lateral (axillary) buds. If the apical bud is removed, lateral buds are likely to grow, but if auxin (say in an agar block or mixed with lanolin) is applied to the cut tip of the shoot, lateral bud growth is again inhibited. Removal of this apical dominance is an important part of the practice of pruning and may, for example, be used to stimulate bushy growth or development of fruit buds (Figure 2.5).
- **Formation of lateral roots** – growth of the primary root is *inhibited* by high auxin concentrations, but auxin *stimulates* initiation of lateral roots above the root hair zone. It will also stimulate the initiation of roots at the bases of cut stems. This is particularly useful in horticulture, where 'hormone rooting powder' is used to encourage root production in stem cuttings.
- **Abscission of leaves and fruit** – IAA delays early stages of leaf abscission but promotes later stages, probably through its stimulation of ethene

Figure 2.5 Pruning of fruit tree: removal of apical buds on these side shoots encourages development of fruit buds. In this young apple tree, summer pruning is used as a means of restricting vigorous growth and to help control the shape of the tree.

QUESTION

Using the information in the text, make a list of the ways in which knowledge of the effect of auxin is used commercially in horticulture and agriculture.

production (page 25). It can be used to delay abscission of fruits or 'fruit drop', ensuring the fruit stays on the tree until harvest.

* **Fruit development** – auxin is produced in pollen and in the developing seed and probably acts as the stimulus for development of the fruit after fertilisation. Treatment of unpollinated flowers with auxin may result in the development of seedless fruits.

* **Synthetic auxins as weedkillers** – synthetic compounds, such as 2,4-D (2,4-dichlorophenoxyacetic acid) are used as selective weedkillers (herbicides). Their use depends on the way plants respond to different concentrations of auxins. Dicotyledons (such as thistles and dandelions) are very sensitive to 2,4-D and show abnormal growth, resulting in distortion of the internodes and rooting system, and then death of the plant. Monocotyledons (such as grasses and cereal crops) are unaffected, so such compounds can be used selectively to remove dicotyledonous weeds from a lawn or from a field with a cereal crop.

Gibberellins

The term **gibberellin** is applied to a group of substances of similar chemical structure, some of which show biological activity. The first evidence of the existence of these plant growth substances came in the 1930s, when some Japanese scientists isolated the substance now known as gibberellic acid (GA_3) from the fungus *Gibberella fujikuroi*. It was well known that rice plants infected with this fungus grew very tall, but did not produce seeds, hence the local name of 'foolish seedling' disease. In the 1950s, the nature of gibberellins and their effects began to be known and studied more widely.

Gibberellins have now been extracted from higher plants, with the highest levels occurring in immature seeds, though they have been detected in other parts of plants. Because of the low concentrations at which they occur, it is difficult to be sure of their natural, or **endogenous**, role in regulation of plant growth. However, a number of effects are now well established for exogenously applied gibberellins and some of these are summarised below. Although there are several known gibberellins, most of the effects to be described can be brought about by the application of gibberellic acid.

* **Stem elongation in dwarf plants** – when treated with gibberellic acid, plants that are genetically dwarf forms show elongation of their stems to become comparable to normal or tall forms. This has been demonstrated in dwarf peas (*Pisum sativum*) and dwarf maize (*Zea mays*). Gibberellic acid has no effect on genetically normal or tall forms. Gibberellins are used in the sugarcane industry to stimulate the growth of the internodes during the winter season, thus increasing the yield.

* **Bolting in long-day plants** – some plants show photoperiodism with respect to their flowering, which occurs as a response to day length (Figure 2.6). Long-day plants remain in a rosette form during short days, but require long days to grow a long flower stalk (bolt). This effect can be overcome by applying gibberellic acid; thus long-day plants can be made to bolt during short days. Similarly, some long-day plants have, in addition, a requirement for a cold period before flowering, and this can also be overcome by the application of gibberellic acid.

* **Fruit development** – gibberellic acid can be used to promote fruit

Figure 2.6 (top) Chinese cabbage (Brassica rapa) as we know it for a vegetable. This is the rosette form which develops during short day-length. (bottom) Longer day length triggers bolting and development of a flower stalk bearing flowers. Gibberellins can be used to induce bolting of long-day plants when the days are short.

development and growth. This application is used commercially to increase the size of seedless grapes. The size and shape of the bunch is also improved because growth of the fruit stalk is stimulated. In citrus fruits, gibberellins are used to delay **senescence** (ageing), allowing the fruits to stay longer on the tree and thus extending the market period.

- **Seed germination** – some seeds have a requirement for light or cold to break dormancy and allow germination. Gibberellic acid can be used to induce germination and thus overcome these requirements.

- **Enzyme production during germination of seeds** – food stored as starch is mobilised or broken down during germination by the action of hydrolytic enzymes (α- and β- amylase). These two enzymes break different linkages in the starch molecule to produce maltose, which can be further broken down to glucose by the action of maltase. In barley and other grains ('seeds') in the grass family, there is evidence that endogenous gibberellins from the embryo stimulate the aleurone layer to release α-amylase. This has important applications in the brewing industry, which utilises germinating barley grains as a source of malt for making beer. After steeping in water, the grains are sprayed with gibberellic acid, which stimulates production of enzymes: first those which break down the cell walls, then release of α-amylase, so increasing the yield of malt before fermentation by yeasts.

- **Gibberellin synthesis inhibitors** – these can be used to counteract the effects described for gibberellins and prevent growth. In commercial situations they are used to keep a compact shape in certain plants, such as chrysanthemums and poinsettias, or to minimise growth of shrub plants along the roadside so that they require less frequent trimming.

Cytokinins

As the name suggests, **cytokinins** are linked to cell division (**cytokinesis**). The discovery of this group of plant growth substances came from attempts to find ways of growing cells in culture, outside the plant, similar to growing cultures of microorganisms. In plants, cells grow and differentiate, but once they have assumed their function, mature cells generally do not divide or show further changes unless stimulated to do so. One example of renewed activity of plant cells is seen in the response to wounding.

Around the beginning of the 20th century, the German scientist Haberlandt believed that individual plant cells had the potential to develop into any type of specialised plant tissue. However, it was not until the 1950s that success was achieved in growing isolated cells in culture media. From these **tissue cultures**, whole plants were produced (Figure 2.7). A typical culture medium contains a source of energy (for example, sucrose), a range of inorganic ions, and other organic nutrients, such as amino acids and some vitamins. In the early stages, the cells being cultured are carrying out heterotrophic nutrition. A key to the success in the culturing of cells was the addition of growth-regulating substances, particularly auxin and cytokinin.

Naturally occurring cytokinins have now been extracted from a wide range of living plant tissues. The first to be identified was **zeatin**, found in the immature endosperm of maize (*Zea mays*). Zeatin in the liquid endosperm of

Figure 2.7 Plant tissue culture on nutrient agar in a Petri dish. Growing cultures can be subcultured repeatedly and stored at low temperatures.

coconut ('coconut milk') is a source that was also used in early cell culture experiments. Zeatin is probably the most commonly occurring cytokinin in plants and has also been found in fungi (yeast) and bacteria. Other synthetic compounds, such as **kinetin** and **6-benzylamino purine** (**BAP**), show similar activity in being able to stimulate cell division.

In the plant, cytokinins are found mainly in young meristems, such as the apices of roots and shoots, where cells are dividing rapidly. They appear to be transported along with water in the xylem, though they have also been detected in the sap from phloem. Cytokinins are involved in the regulation of various processes in plants, including development (morphogenesis) of roots and shoots, development of lateral buds, cell enlargement, maturation of chloroplasts and a contribution to the processes that delay senescence. Their activity is linked to events in the cell cycle, in particular to nucleic acid metabolism and protein synthesis, though the mechanism of action is not known. The effects of cytokinins are often dependent on other growth substances, such as auxin, and may affect tissues differently at different stages of growth. As with gibberellins, evidence for cytokinin activities comes mainly from exogenous application to excised (cut) sections of plant material.

Abscisic acid

The effects of **abscisic acid** (**ABA**) are generally to inhibit or alter growth of plants, and are often linked with conditions of environmental stress. Abscisic acid was first associated with **abscission** (cutting off) of leaves at leaf fall and of ripe fruits. Since its chemical identification in the 1960s, it is now known that ABA is involved in a wider range of plant processes, including dormancy in seeds and buds and closure of stomata in times of water deficit. Abscisic acid has been detected in organs throughout the plant and is transported in both the xylem and phloem.

Dormancy in seeds and buds of woody plants is an important adaptation for survival through unfavourable conditions, particularly cold temperatures. Concentration of ABA is usually higher in dormant seeds compared with seeds that are not dormant. It is probably the balance between ABA as an inhibitor and cytokinins and gibberellins as growth promoters which is responsible for the transition from dormancy to germination. One effect of ABA may be to inhibit the synthesis of hydrolytic enzymes which allow the breakdown of storage reserves in the seed, whereas gibberellic acid (GA_3) appears to stimulate the release of α-amylase from the aleurone layer.

ABA also appears to inhibit **growth** which is promoted by auxin and here the mechanism of action may be to block H^+ secretion, thus interfering with the loosening of the cell wall which would allow elongation of the cells (see page 21).

Evidence for a link between ABA and **environmental stress** in a plant comes from the marked increase in concentrations of ABA that can occur in a plant subjected, for example, to drought conditions. Application of ABA can induce rapid closure of stomata. Within the leaf, redistribution of ABA from the chloroplasts (where it accumulates) to the guard cells leads to change in turgor, resulting in closure of stomata and thus reducing loss of water from the leaves. Plants may also lose their leaves (leaf abscission) as a response to water stress.

While ABA was originally associated with **abscission**, in most plants ethene may be the more important hormone, whereas ABA is clearly linked with senescence (ageing). Segments of leaves, for example, turn yellow due to breakdown of chlorophyll and this process is accelerated by the application of ABA, whereas cytokinins antagonise the action of ABA and delay senescence.

Ethene (ethylene)

Ethene is the only gas known to act as a plant growth substance. Its effects were first recognised around the turn of the 20th century, when it was noticed that trees close to coal-gas street lamps lost their leaves more than other trees, that bananas ripened prematurely when packed alongside oranges, and lemons would turn yellow when kept near heaters burning kerosene. Although the gas ethene was identified as being present in these situations, it was not until the 1950s that its importance in the metabolism of mature plants began to be understood.

Ethene is found widely in all parts of plants but at low concentrations. Because it is a gas, its distribution is by diffusion rather than direct transport. It can show a marked rise in some fruits (such as tomatoes, apples and avocados) during the ripening process. This is associated with a similar rise in respiration rate, measured by carbon dioxide output. Ethene is synthesised from the amino acid methionine, and it appears that the final step in the pathway is critical for controlling the concentrations of ethene present. A number of factors affect this final step, including environmental stress (drought, flooding, chilling or pollution), wounding or other plant growth substances. Ethene is known to be involved in certain responses to stress, such as abscission and healing of wounds, as well as the ripening process. Auxin appears to stimulate ethene synthesis.

The process of **abscission** in leaves, flowers or fruits is linked to the differentiation of a layer of cells known as the abscission layer. The cell walls become weakened due to the activity of enzymes, such as cellulase acting on cellulose and polygalacturonase (PG) on pectic substances. Ethene appears to promote abscission, whereas auxin prevents it. Part of the ripening process in fruits involves similar softening of tissues, as illustrated by the change from firmness to mushiness in ripening tomatoes.

Ethene has a number of important commercial applications. (See *Respiration and Coordination*, Chapter 9.)

Synergism and antagonism

Many aspects of plant development, response and growth are influenced by more than one growth substance. If two or more growth substances interact with each other, synergism occurs and the effects of individual substances may be enhanced. Auxins and gibberellic acid augment cell elongation in stems. Gibberellic acid reinforces the action of auxins in maintaining apical dominance in terminal buds. In the later stages of leaf abscission, auxin acts synergistically with ethene in bringing about leaf fall.

If plant growth substances oppose each other's effects, they are said to show antagonism. Cytokinins are antagonistic to the action of auxins in promoting apical dominance. Abscisic acid is antagonistic to auxin action in cell elongation in stems. In the early stages of abscission, auxin is antagonistic to ethene.

| PRACTICAL | **Investigating the effect of weedkillers on plant growth** |

Introduction

Synthetic auxins, such as 2,4-D (2,4-dichlorophenoxyacetic acid) and MCPA (4-chloro-2-methylphenoxyacetic acid) are used as **selective herbicides**, because they kill broad-leaved (dicotyledonous) weeds, leaving the narrow-leaved (monocotyledonous) plants unaffected. For this reason, synthetic auxins are particularly useful for weed control in cereal crops, including barley and maize. The aim of this experiment is to show the effect of selective herbicides on the growth and development of plants. For the purposes of demonstration, plastic trays containing barley (*Hordeum* sp.) and radish (*Raphanus* sp.) plants can be used – radish is similar to some of the weeds which grow in cereal crops.

Materials

- Plastic trays containing barley and radish plants, growing in a suitable compost
- Hand sprayer containing a selective herbicide, such as Verdone® (lawn weedkiller, obtainable from garden centres)

Method

1 Follow the instructions carefully and spray one tray of plants with the herbicide. Leave the other tray untreated as a control.

2 After 48 hours, record any changes in appearance of the plants.
3 Continue your observations, at weekly intervals, for 4 to 6 weeks.

Results and discussion

1 Record your observations in a suitable table.
2 What alternatives are there to the use of chemical herbicides?

Further work

1 You could investigate the effects of herbicides using other combinations of monocotyledonous and dicotyledonous plants.
2 Plan and carry out an investigation into the effects of weed competition, and treatment with a herbicide, on the yield of the crop plant.

SAFETY
When using the spray:
- Keep off skin
- Do not breathe spray
- Wash off any splashes
- Wash hands and exposed skin after use

3

Classification, distribution and succession

Classification

An understanding of classification and the principles on which it is based is of great importance in studying and describing the distribution of organisms in their habitats. Classification enables us to link organisms with similar characteristics into groups for easy reference, so it is important that a suitable, sensible system for naming organisms exists and that such a system should be recognised internationally. The use of common names for organisms, such as 'frog', 'buttercup' and 'mushroom', may be widely used in Britain, but these names lack precision. There are many different types of frogs and buttercups and the term 'mushroom' is commonly used for any edible member of the group Basidiomycota.

In order to avoid confusion, the **binomial system of nomenclature**, based on a scheme devised in the 18th century by the Swedish naturalist Carl von Linné (otherwise known as **Linnaeus**), is used. In this system, each organism is given a name consisting of two parts:
- a **generic** name, which states the **genus** and is common to a group of closely related organisms
- a **specific** name, stating the **species**, which is unique to a particular organism and often descriptive of one of its characteristics.

Both names are given in Latin, the generic name beginning with a capital letter and the specific with a lower case letter. It is the convention to use italics when printing the names, but when handwritten it is usual to underline them. For example, in a particular habitat, we might find *Ranunculus bulbosus* (the bulbous buttercup), together with *Primula vulgaris* (the common primrose), *Lumbricus terrestris* (the earthworm) and *Capsella bursa-pastoris* (the shepherd's purse). Latin was used by Linnaeus because it was understood by most educated people in the 18th century and its use persists in modern times. It enables world-wide recognition of names, and any new species which are discovered must be named according to this binomial system, with a detailed written description of the organism.

Many different schemes of classification have been devised, from Aristotle in the 3rd century BC, dividing animals into those with red blood and those without, to the proposal by the American biologist R. H. Whittaker in 1959 of the **Five Kingdom System**. The currently recommended scheme is based on Margulis and Schwartz's modification of Whittaker's proposals and recognises the following five kingdoms of living organisms:
- **Prokaryotae**
- **Protoctista**
- **Fungi**
- **Plantae**
- **Animalia**.

The main distinguishing features of these kingdoms are outlined in Table 3.1, together with examples of the groups included in them.

Table 3.1 *The Five Kingdoms*

Kingdom	Characteristics	Representative groups
Prokaryotae	organisms lack nuclei organised within membranes; lack envelope-bound organelles; lack 9 + 2 microtubules	non-photosynthetic: *Escherichia coli,* photosynthetic: cyanobacteria *Anabaena*
Protoctista	eukaryotic organisms with organised envelope-bound nuclei which are neither fungi, plants nor animals; often unicells or assemblages of similar cells	green algae: Chlorophyta brown algae: Phaeophyta Protozoa
Fungi	non-photosynthetic eukaryotic organisms with a protective, non-cellulose wall; absorptive methods of nutrition; usually multinucleate hyphae; spores without flagella	Zygomycota: *Mucor* Ascomycota: *Saccharomyces* Basidiomycota: *Agaricus*
Plantae	multicellular, photosynthetic, eukaryotic organisms; cell walls contain cellulose	Bryophyta: mosses Filicinophyta: ferns Angiospermophyta: flowering plants
Animalia	multicellular, non-photosynthetic eukaryotic organisms with nervous coordination	Cnidarians Annelids Arthropods Molluscs Chordates

Taxonomy

Taxonomy is the study of the classification of groups of organisms. These groups are called **taxa** (sing. **taxon**), and are organised into hierarchies, which attempt to take into account their supposed evolutionary descent. A taxon is a group of organisms which share common features. There are seven major levels of taxon, of which the highest is the kingdom and the lowest is the species.

A **species** is defined as a group of individuals with a large number of features in common. Members of a species can interbreed and produce fertile offspring, but members of one species do not normally breed with members of another species. This definition is restricted to sexually reproducing organisms and cannot apply where reproductive behaviour has not been observed (as is the case with fossils), organisms that only reproduce asexually, or parthenogenetic forms. A slightly different definition, which specifies a group of organisms showing a close similarity in morphological, biochemical, ecological and life history characters, is more generally used and nearly all species are established on this basis. Some species are widely distributed and show local or regional differences, which justify recognition and result in the taxon being split into a

> **DEFINITION**
>
> A **species** is a group of organisms which can interbreed to produce fertile offspring.

number of **subspecies**, **varieties** or **races**. Such differences, although small, are thought to be significant, but do not prevent interbreeding.

Between the lowest taxon (the species) and the highest (the kingdom) there are five other taxa: **genus**, **family**, **order**, **class** and **phylum**. The relationships between these taxa are summarised in Table 3.2. As the classification proceeds from the kingdom to the species, the number of organisms in each taxon decreases, but the number of shared features increases. Some taxa, such as the phylum **Chordata**, are very large and have been subdivided into subphyla for convenience.

Table 3.2 *Taxa used in the classification of organisms*

| Taxon | Description | Organism | | |
		Bladderwrack	Garlic	Human
kingdom	largest group of organisms sharing common features	Protoctista	Plantae	Animalia
phylum	major subdivision of a kingdom	Phaeophyta	Angiospermophyta	Chordata
class	a group of related orders; subdivision of a phylum	Phaeophyceae	Monocotyledonae	Mammalia
order	a group of related families; subdivision of a class	Fucales	Liliales	Primates
family	a group of closely related genera; subdivision of an order	Fucaceae	Liliaceae	Hominidae
genus	a group of related species; subdivision of a family	*Fucus*	*Allium*	*Homo*
species	a group of organisms capable of interbreeding and producing fertile offspring	*F. vesiculosus*	*A. sativum*	*H. sapiens*

In any study of living organisms in their habitats, it is useful to be able to place them in their taxonomic groups. For this to be achieved with any accuracy, a knowledge of the distinguishing external features of the different taxa is required and, for correct identification to genus and species level, it is necessary to use **keys**.

Distribution of plants and animals: effects of biotic and abiotic factors

A number of physical factors affect the distribution of organisms in their habitats. These physical factors are often referred to as **abiotic**, to distinguish them from the **biotic** factors, which involve the effects of other living organisms, including humans, on the distribution of species. Some physical factors, such as light, have widespread effects and are important in a range of different situations, both aquatic and terrestrial. Others, such as wave action, are only of significance in aquatic environments.

The physical factors can be divided into:
- **climatic** – temperature, light, wind and water availability
- **soil** – often referred to as **edaphic** factors
- **topographic** – altitude, aspect (whether north-facing or south-facing) and inclination (steepness of slope)
- others – such as wave action, which are relevant in specific situations.

Temperature

Most living organisms have an optimum temperature range within which they can survive; thus variations in temperature will affect the rate at which they grow. The main source of heat is the sun's radiation and the temperature of a habitat will depend on its latitude, the season of the year, the time of day and its aspect. The temperature of any environment can also be affected by the presence or absence of vegetation.

Temperatures below freezing will cause physical damage to living cells as ice crystals will form. If the temperature becomes too high, then enzymes are denatured, causing growth and metabolism to be disrupted. Aquatic habitats tend to have more stable temperature conditions than terrestrial ones, due to the high specific heat of water. It takes a great deal of heat energy to bring about a significant increase in the temperature of a large body of water, so there is less fluctuation in the temperature of an aquatic habitat than there is in a terrestrial one.

Measuring temperature

Temperature is an important parameter as it can affect the metabolic activities of many organisms and, in turn, their growth or reproductive rates. The solubility of oxygen in water is inversely related to the temperature, so this is an important factor to consider in aquatic ecosystems. The ordinary glass thermometer is not really suitable for field work as it is too fragile. There are a number of electronic thermometers available, which have several advantages. For example, if they have a temperature probe, it is possible to measure the temperature of water at different depths, or the probe can be pushed into soil to measure soil temperature. Some temperature probes are ideally suited to interfacing with datalogging equipment, which makes it possible to make continuous recordings over extended periods of time.

Light

Light is the source of energy for photosynthesis, so it influences primary productivity. As the consumers are directly or indirectly dependent on the primary producers, light is therefore essential for the maintenance of life. Light has a number of other roles, one of which, the length of the photoperiod, has considerable influence on the behaviour of both plants and animals. Many activities, such as flowering and germination in plants and reproductive behaviour in animals, are linked to the photoperiod so that they can be synchronised with the seasons.

The need for light by plants affects the structure of communities. A clear example of this is seen in the different layers of vegetation in a woodland, where shade-tolerant species are found underneath the tree canopy. In aquatic habitats, the plants are confined to the surface of the water or to the shallow water at the margins. Even in shallow water, the penetration of light for photosynthesis may be affected by the turbidity due to suspended particles.

Measuring light intensity

Measurement of light intensity is not particularly easy. Some light meters will give an indication of the relative light intensity and, unless the instrument is calibrated, will give readings in arbitrary units. This is adequate if you are recording, for example, the change in light intensity as you move into woodland from an area of open ground. When taking readings, it is important that the solar cell is always pointing in the same direction, usually towards the Sun, and that readings are taken at the same height above ground. To make comparisons meaningful, it is important that the light source remains constant while you are taking the readings. If the Sun becomes hidden by a cloud, it will obviously change the reading in one place. Recordings should, therefore, be made as quickly as possible. Light sensors are also available which make continuous recording possible over a period of time.

Wind and water movements

Air movements, or wind, are of significance in the distribution of organisms as they interact with other physical factors. This is particularly apparent in exposed coastal or upland areas where the prevailing wind may affect the growth of trees and shrubs. Air movements accelerate the rate of evaporation of water and hence affect transpiration. Increased wind speed increases the likelihood of freezing, so that the buds on the more exposed side of a tree are more likely to suffer frost damage. Some seed and spore dispersal mechanisms rely on air movements, and strong winds can affect the migration of birds.

Water movements include currents and the ebb and flow of tides, the latter possibly interacting with air movements to bring about wave action. Any movement of water will have an eroding action on soil and rocks, as well as moving living organisms from one place to another, unless they are rooted or attached to a substratum. The churning action of water movements results in aeration of the water and it may also contribute to the turbidity. Organisms in aquatic habitats show a wide range of adaptations to life in water.

Rate of flow in streams or rivers is an important parameter because of its influence on the organisms inhabiting the water. As the current increases, organisms which are unable to swim against it, or to take hold, are likely to be washed away. Faster flowing water is likely to be better oxygenated than sluggish or still water, because of the mixing effect.

Measuring wind and water movements

Several hand-held meters are available that can be used in the field. Examples are wind-speed indicators, and flow meters, including the stream-flow meter which has the advantage of being able to be used at specific depths. Flow meters are generally expensive and, as an alternative, adequate measurements of flow rate can be made using the timed float method (see *Practical: Timed float method to determine current velocity*).

Water availability

Plants vary widely in their ability to tolerate a shortage of water. **Hydrophytes** are adapted to living in waterlogged or submerged conditions, whereas **xerophytes** show adaptations which reduce water loss and can survive in conditions where water is scarce. **Mesophytes** are those plants which thrive best in conditions where there is an adequate supply of water, so that water loss through transpiration can usually be replaced by water uptake from the soil. In temperate regions, these plants do show some adaptations to seasonal variations in water availability: woody plants may shed their leaves in autumn and the aerial parts of herbaceous perennials may die down, thus reducing water loss during the unfavourable season.

Many terrestrial animals show adaptations which enable them to regulate water loss and thus survive in conditions of water shortage. Emphasis is on the conservation of water because, in areas where water is in short supply, the rate of evaporation is also high. Aquatic animals also need to be able to maintain a constant internal environment, so must regulate water uptake and water loss. Those animals living in fresh water, where the concentration of body fluids is often higher than the concentration of ions in their environment, have a tendency to take up water by osmosis. They show adaptations which prevent or reduce water uptake and also possess a means of getting rid of the excess water. In marine habitats, the concentration of ions in sea water is higher than that in fresh water, so animals may show adaptations to counteract the tendency to lose water.

Salinity

Plants which can tolerate high levels of salt are referred to as **halophytes**, and are typically found growing in estuaries and saltmarshes, where their roots may be immersed in sea water. The degree of salinity to which they are exposed can vary, depending on their location and the tides. Salinity levels can be lower than sea water in a tidal estuary, but higher in a saltmarsh due to the evaporation of water from the soil at low tides. Adaptations to these changing conditions often involve the maintenance of high salt concentrations in the tissues or the development of special tissues in which water can be stored.

Some animals can adapt to changes in salinity, but as many animals are motile they will generally move to an area where they are best suited to the prevailing conditions.

Measuring salinity in aquatic habitats

Salinity is a measure of the salt content of sea water. The salts in the sea are mainly sodium and chloride, but smaller concentrations of other ions are also present. These include potassium, magnesium, calcium and sulphate ions. The

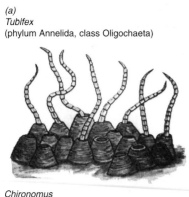

(a)
Tubifex
(phylum Annelida, class Oligochaeta)

Chironomus
(phylum Arthropoda, class Insecta)

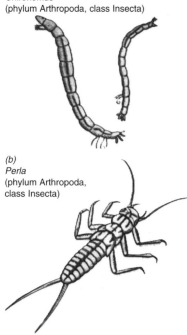

(b)
Perla
(phylum Arthropoda,
class Insecta)

Figure 3.1 Organisms typically found in: (a) poorly oxygenated fresh water; (b) well-oxygenated fresh water

Table 3.3 *Oxygen content of water saturated with air at normal pressure*

Temperature / °C	Dissolved oxygen / mg dm^{-3}
0	14.66
5	12.37
10	10.92
15	9.76
20	8.84
25	8.11

salt concentration is usually given the symbol ‰ (parts per thousand). Salinity can be determined by titrating a sample of sea water against silver nitrate solution.

Oxygen concentration

Oxygen is of vital importance to aerobic organisms in any habitat. The oxygen concentration of the atmosphere is more or less constant at around 21 per cent, but that of the soil atmosphere is slightly lower due to respiration of soil organisms. The oxygen concentration in aquatic habitats can vary greatly. It may be very low in still, undisturbed water, with anaerobic conditions in the mud at the bottom. Any disturbance of the water will bring about aeration, as will the presence of actively photosynthesising plants. An increase in temperature will reduce the amount of dissolved oxygen in water and could result in a reduction in the numbers of organisms present.

Most living organisms are unable to tolerate anaerobic conditions and will die quickly if the level of oxygen falls. One of the effects of organic pollution in fresh water is a reduction in dissolved oxygen levels as a consequence of the activity of aerobic bacteria. Some invertebrates are more tolerant to low levels of oxygen than others and can survive in poorly oxygenated water. These tolerant organisms include *Tubifex* worms and the larvae of certain midges, notably *Chironomus*. These organisms both contain haemoglobin which helps them to obtain oxygen. Larvae of stoneflies, such as *Perla* spp., will only live in well-oxygenated water, which is characteristic of fast-flowing streams. These organisms are illustrated in Figure 3.1.

For more information on the effects of oxygen concentration on aquatic animals, see *Exchange and Transport, Energy and Ecosystems*, Chapter 3.

Measuring dissolved oxygen

The solubility of oxygen in water is relatively low and varies inversely with the temperature, as shown in Table 3.3. The amount of oxygen dissolved in water is usually measured either in mg dm^{-3} or as a percentage of air saturation (the amount of oxygen present expressed as a percentage of the amount dissolved in water at equilibrium with air at the same temperature).

There are two main approaches to the measurement of dissolved oxygen, using either a dissolved oxygen meter and suitable electrode or chemical determination. Dissolved-oxygen meters are expensive but convenient to use in field work. Chemical analysis is simple and relatively inexpensive.

Soil

Soil covers a large part of the land surface of the Earth and consists of:
- rock, or mineral particles, derived from the breaking up or weathering of rocks (46 to 60 per cent)
- organic material called humus (about 10 per cent)
- water (25 to 35 per cent)
- air (15 to 25 per cent)
- living organisms, such as bacteria, fungi, protozoa, insect larvae, earthworms and moles in variable numbers depending on the soil type and location.

The mineral particles vary in size from gravel, with particles greater than 2 mm in diameter, to clay, with particles less than 0.002 mm in diameter. Table 3.4 categorises soil particles according to size. The proportions of the different-sized particles determine the texture and properties of a soil. These proportions can easily be determined by shaking up a sample of soil with water in a measuring cylinder, allowing it to settle and then estimating the volume of different-sized particles as they appear in the sample. The larger, heavier particles sediment out more quickly than the lighter, smaller ones. Clay particles may remain in suspension but can be made to settle out by the addition of calcium hydroxide, which flocculates the particles. Any organic material usually floats on the top. The results of this technique are illustrated in Figure 3.2.

Water drains more easily through a soil with a high proportion of larger sized particles, such as a sandy soil, than it does through a clay soil, where the smaller sized particles predominate. Clay soils can become waterlogged if rainfall is high, though they will retain water in drier conditions. Sandy soils do not retain much water and the rapid drainage results in the loss of mineral ions through leaching.

The organic matter present in the soil is an important source of food for the decomposers and detritivores. These organisms bring about the breakdown of the dead remains of plants and animals, releasing mineral ions into the soil, which are then available for uptake by plants.

Table 3.4 *Table of mineral particle sizes in soils*

Particle type	Diameter / mm
clay	<0.002
silt	0.002–0.02
fine sand	0.02–0.2
coarse sand	0.2–2.0
gravel (small stones)	>2.0

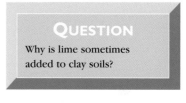

QUESTION

Why is lime sometimes added to clay soils?

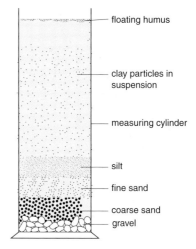

Figure 3.2 Sedimentation to show soil composition

EXTENSION MATERIAL

More details on soil structure and soil types

Soil type can have an effect on the habitat, because it will determine the type of vegetation that is present, which in turn will determine the number and species of animals. Soil type can be determined by digging a **soil profile** to expose the different layers. Figure 3.3 illustrates two different soil profiles. A soil profile usually shows the following characteristic layers, or **horizons**:

- **A horizon** – the topsoil, from which minerals tend to be removed by leaching; this layer is often subdivided into: A_0, the litter layer; A_1, the humus layer; and A_2, the leached layer
- **B horizon** – the subsoil, which tends to collect the minerals leached out of the A horizon
- **C horizon** – consisting of weathered parent material
- **D horizon** – the parent rock, or bedrock.

Brown earths, or brown forest soils, are associated with temperate deciduous woodland at low altitudes. The profile is a simple one, with a relatively thin litter layer and A and B horizons of a similar brown colour. These soils tend to have a pH range of 4.5 to 8.0 and support large numbers of soil organisms which break down the litter into humus. There are usually earthworms present which pull litter down into the soil as well as mixing the soil layers in the course of their burrowing activities.

Podzols are typically associated with coniferous woodland and heathland at higher altitudes in temperate regions. The profile is more complicated than that of a brown earth, with a much deeper litter layer, an ash-grey A_2 horizon and an 'iron pan' in the B horizon. The litter consists of leaves from the pine trees, which take a long time to break down. The pH range of these soils is in the range of 3.0 to 6.5, so there are fewer soil microorganisms and earthworms are scarce. Most of the breakdown of the litter is achieved by fungal action. The soils are sandy and occur in regions of high rainfall, so a great deal of leaching occurs, leaving the A_2 horizon an ash-grey colour. The minerals, including iron, and humus accumulate in the B horizon and give rise to a dark layer called the 'iron pan', which is hard and impenetrable to plant roots.

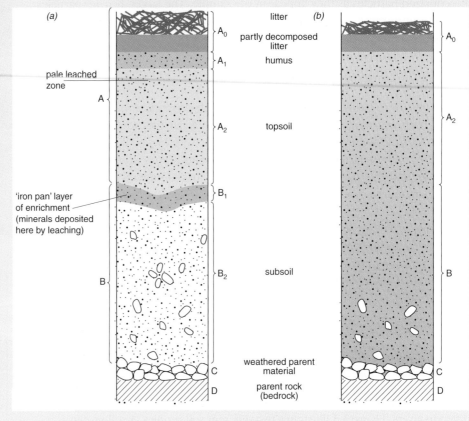

Figure 3.3 Soil profiles of: (a) a podzol; and (b) a brown earth

Investigating soil type

Soil type can be investigated by sedimentation and the digging of soil profiles. Different soils can be compared and contrasted by these methods.

Investigating soil water

Soil water is divided into several categories. After heavy rain, before water has had a chance to drain away, the soil is said to be at **saturation**. Much of this water (referred to as the **gravitational water**) will move down through the spaces between soil particles, but some will remain adhering to soil particles. This is known as **capillary water**. Soil that contains all the capillary water it can hold against gravity is said to be at **field capacity**. This can be determined by placing a soil sample in a funnel, adding water and leaving it until water stops dripping out. The percentage of water in the soil sample can then be determined (see *Practical: Determining the water content of a soil sample*, page 46).

pH

The pH value is a measure of the hydrogen ion concentration of an aqueous solution and indicates the level of acidity or alkalinity. The pH value of a soil has an effect on the availability of mineral ions and, to some extent, can determine the type of vegetation which will grow. Certain plants, such as heather (*Erica* sp.), grow well in the acid conditions of heathland, while others, such as dog's mercury (*Mercurialis perennis*), will only grow in more alkaline conditions. The type of vegetation in an area will have an influence on the animal populations. In Britain, soil pH values range from pH 4.0, which is acid, to pH 8.0, which is slightly alkaline. A pH of 7.0 is neutral. Many plants appear to be sensitive to the presence or absence of calcium ions in the soil.

The pH of aquatic environments can cause variations in the distribution of organisms. The range of pH in lakes and ponds is 4.7 to 8.5, and though some aquatic species, such as *Gammarus* (the freshwater shrimp), can tolerate a wide range, others are only found in either acid or alkaline conditions. The presence or absence of calcium ions also influences the distribution of members of the Mollusca and Crustacea. Both these groups need calcium ions, the molluscs for their shells and the crustaceans for their integuments, so both are restricted to situations where calcium is present.

Measuring soil pH

The pH of soil sample may be determined conveniently using universal indicator solution. Place about 1 cm depth of soil in a test tube and add 1 cm of barium sulphate. Barium sulphate helps clay particles to settle, leaving a clear solution. Add 10 cm^3 of distilled water and shake the tube thoroughly, then add a few drops of universal indicator solution. Compare the colour with the colour chart provided with the indicator solution, and record the pH value. As alternatives to indicator solution, pH papers may also be used, or a pH electrode and meter, to find the pH of the soil solution. (See *Practical: Determining the pH of a soil sample*, page 46.)

Qualitative and quantitative field techniques

Before undertaking an ecological survey, it is necessary to consider the factors which we want to measure. Within ecology, studies are of two broad types: **autecology** and **synecology**. Autecology considers the ecology of a single species, such as the distribution of limpets on a rocky shore, whereas synecology focuses on organisms and their environment. In this case, the whole of the rocky-shore community would be investigated.

In an ecological investigation, careful planning is required to identify the sorts of biotic and abiotic factors which are important in a particular ecosystem. Measurement of these factors will, of course, depend on the resources available, but useful data can be obtained with simple equipment. It is the design of the investigation which is important, rather than the amount of elaborate equipment which is used. Figure 3.4 illustrates a variety of simple equipment used in ecological studies and more details can be found in *Tools, Techniques and Assessment*, Chapter 6.

ADDITIONAL MATERIAL

Plants on calcareous soil
Plants which can thrive on calcareous soils, where calcium ions are abundant, are termed calcicoles, or calcicolous. Plants which cannot tolerate calcium are termed calcifuges and their presence indicates the absence of calcium in the soil. When these species are grown on their own in the 'wrong' type of soil, they still thrive, so it is likely that some other factor, such as the availability of other mineral ions, climatic conditions or competition, is involved.

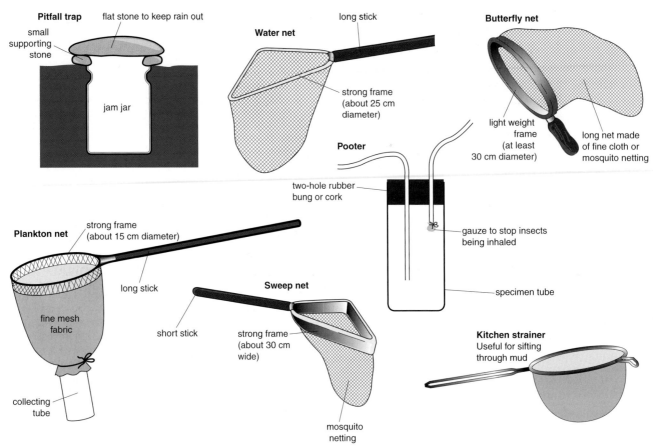

Figure 3.4 Examples of simple equipment used in ecological studies

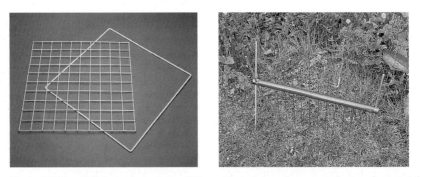

Figure 3.5 A gridded quadrat, giving 100 sample points, a frame quadrat and a point frame. A point frame is used to determine the percentage cover of vegetation.

Random sampling

Suppose we want to determine the number of dandelions (*Taraxacum officinale*) present in a field, or the number of limpets (*Patella vulgata*) on a rocky shore. It would obviously be impractical to count every individual, so instead we need to use a sampling technique. In a study of this sort, random sampling with a quadrat frame is used (Figure 3.5). A quadrat frame is usually made of wood or metal and is used to take a sample of the area under investigation. It is assumed that the area within the quadrat frame is representative of the entire area. There are three questions which need to be considered when using a quadrat is this way.

- How should the quadrat be positioned?
- What size quadrat should be used?
- How many samples should be taken?

Positioning the quadrat

We adopt a scientific approach to obtaining a number of random samples by using pairs of random coordinates. Random coordinates are used to position the quadrat using a fixed reference point, for example, one corner of a field being studied. Two tape measures can be set out at right angles to each other to set up a pair of axes. Pairs of random numbers, for example, 5 and 13, are then used to indicate the directions in the x and y axes; in this case it would be 5 m along in one direction, then 13 m along perpendicular to that axis.

Random numbers can be obtained in several ways. Numbers representing the distances in metres can be written on separate pieces of paper, which are placed in a small bag or box and shaken well; then one number is drawn out (the 'numbers out of a hat' method). Each time a number is drawn out, it should be noted, replaced, and the procedure repeated. Tables of random numbers are available in books of statistical tables or ecological techniques. Two adjacent columns can be used as the x and y coordinates. A computer (and some calculators) can be used to generate random numbers.

Quadrat size

Quadrats of various sizes are available, for example 0.25 m² and 1 m². In general, smaller quadrats are more reliable than a large quadrat, but there are practical considerations which need to be taken into account. The shape of the quadrat is another factor you could consider. Quadrats are normally square, but does the shape of the quadrat have any effect on estimations of species density? This could be investigated by using quadrats of different shapes, for example, square, rectangular and circular, but having the same area in each case.

Sample size

The distribution of many species is aggregated, that is, they are not uniformly distributed. One sample, therefore, may be completely unrepresentative of the entire population and a calculation of the density based on that sample would be inaccurate. To make the results statistically significant, a large number of samples should be taken, but identifying and counting all the species present in a very large sample can be too time-consuming. We therefore need a method to investigate the effect of sample size on the estimate of density. This can be done by working out the running mean of the density with each sample taken. To illustrate this method, suppose we were finding the density of specimens of common periwinkles (*Littorina littorea*) on a rocky shore, using a 0.5 m² quadrat. The results might look something like Table 3.5.

Initially, the estimated population density fluctuates greatly, indicating that there are too few samples for a reliable estimation of the density. Eventually, the fluctuations decrease and the estimated density will remain more or less the same with each successive sample. The actual number of samples required will depend on the distribution of organisms, but for practical purposes, a minimum of 20 samples should usually be taken. The effect of sample size on population density estimates could form the basis of an individual study.

Table 3.5 *Density of* Littorina littorea *on a rocky shore*

Sample no.	Specimens per quadrat	Density / m^{-2}	Cumulative mean density / m^{-2}
1	7	14	
2	2	4	$(14 + 4) \div 2 = 9$
3	12	24	$(14 + 4 + 24) \div 3 = 14$
4	8	16	$(14 + 4 + 24 + 16) \div 4 = 14.5$
5, etc.	15	30	$(14 + 4 + 24 + 16 + 30) \div 5 = 17.6$

Transects

A transect is a form of systematic sampling, where the samples are arranged in a linear sequence. Transects are particularly useful for recording changes in populations of plants where some sort of transition exists, for example, from the edge of an open field into an adjacent woodland, or across a saltmarsh from the high water mark to the low water mark.

A belt transect is a strip, usually 0.5 m in width, located across a study area in such a way that the transitions in plant populations are highlighted. A tape measure is laid across the sampling area and a 0.25 m^2 quadrat frame laid down at 0.5 m intervals alongside the tape measure. Animals and plants within each quadrat are counted, identified and recorded. With a transect over about 15 m in length, this procedure becomes too time-consuming and it is more usual to carry out quadrat sampling at 1 m intervals.

Succession

The development of a stable community, or **climax community**, takes place in a number of stages and long-term changes in the composition of a community are known as **succession.** Succession occurs because the activities of living organisms over a period of time have a modifying effect on the nature of the environment. Very few large plants can grow where there is no organic matter in the soil, such as on the bare rock of newly erupted islands, sand dunes or glaciated surfaces. Such areas can be colonised by algae and lichens, forming a **pioneer community**. Their activities result in the accumulation of organic debris which, together with the weathering of the rock particles, leads to the formation of a soil in which other organisms can live. Once a soil has been formed, mosses, ferns and small herbaceous plants can grow, replacing the pioneer organisms. Eventually, these smaller plants may be replaced by larger plants until the climax community, a woodland or forest, is formed. Such a succession from bare rock to forest is called a **sere** and each different community in the succession is called a **seral stage**, or **seral community**. This type of succession, from bare rock to woodland, is known as a **primary succession** (Figure 3.6). Where a succession occurs on an area of burnt heathland or cleared woodland which has previously had vegetation on it, then

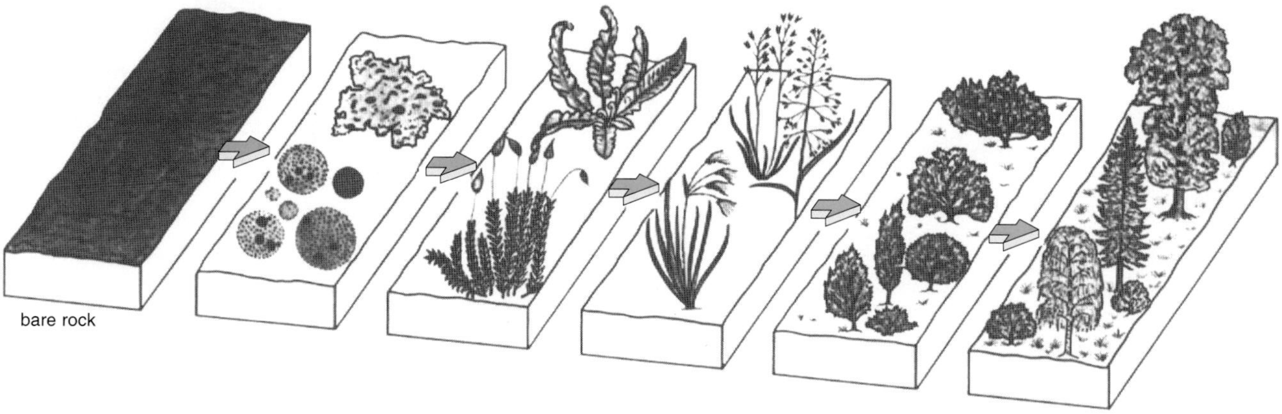

major environmental   lichens and algae   mosses and ferns   grasses (grassland)   shrubs (scrub)   trees
disturbance   woodland / forest

bare rock

Figure 3.6 Diagram showing primary succession from bare ground to woodland

the succession is called a **secondary succession**. In both types of succession, the plant and animal species of neighbouring areas will have an influence on the composition of the flora and fauna. Spores and seeds can be carried by air movements and animals can move independently, both over quite large distances. Larger animals, such as mammals and birds, are amongst the later residents of communities, though they may visit an area undergoing colonisation to seek food. It is unlikely that the conditions will be suitable for them to build nests or find shelter until shrubs and trees have colonised the area. It may take hundreds of years for a primary succession to reach the climax community, but less time for a secondary succession, as the soil has already been formed.

Some general points can be made about the changes taking place during succession.

- The types of animals and plants differ in their characteristics from one seral stage to another. In the early stages, the plants will be annuals, completing their life cycles in one growing season, succeeded later by herbaceous perennials and later still by woody perennials.
- The succession is always associated with increase in biomass. As the soil becomes deeper, it can support a greater amount of vegetation and hence larger numbers of animals.
- The diversity of plant and animal species tends to increase. The greatest number of animal species will be present when the climax has been reached. The number of plant species may be greater before the climax vegetation is reached. This is particularly noticeable where the climax is a beech wood; the dense canopy reduces the number of ground species due to the availability of light.

When conditions are dry, the initial colonisers must be adapted to survive where water is scarce, so tend to be xerophytes. Such a succession is known as a **xerosere**. As succession proceeds, the conditions change and, with the development of an organic soil, more water retention is possible, enabling mesophytes to grow. If the succession starts in shallow water, then a **hydrosere** is formed. In this case, the accumulation of organic debris results in the conditions becoming gradually drier and more favourable to mesophytes. These types of succession are illustrated in Figure 3.7.

CLASSIFICATION, DISTRIBUTION AND SUCCESSION

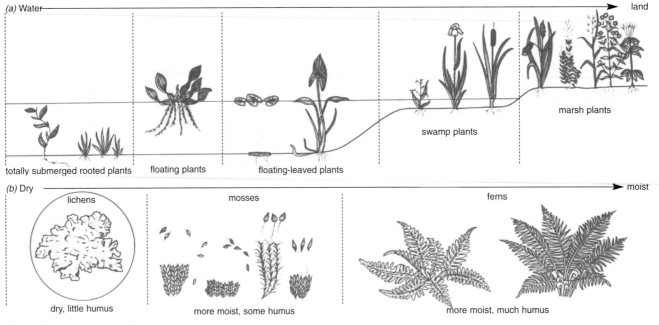

Figure 3.7 (a) Hydrosere; (b) xerosere succession

A climax community is stable and in equilibrium with the climatic conditions. For these reasons it is known as a **climatic climax**. In most of Britain, where the climate is cool temperate, the climatic climax is deciduous (broad-leaved) woodland. Humans and other animals can interfere with this climatic climax and a different equilibrium may be reached, known as a **biotic climax**. Any form of management of the land or aquatic environment can produce a biotic climax. One of the most familiar effects is that of intensive grazing by sheep or rabbits. The growth of shrubs and trees is prevented because the grazing animals feed on the young shoots and the plants never have a chance of becoming established. In this case the biotic climax is grassland. It is seen extensively in the North and South Downs, where sheep are farmed and where rabbits are plentiful. When most of the rabbit population was wiped out due to myxomatosis, large areas of chalk grassland quickly developed into scrub, with hawthorn and other shrubs. Conservationists became concerned as the distinctive chalk flora was in danger of being lost in some areas, so the scrub was cut back in order to preserve the diversity of species.

The type of biotic climax illustrated by chalk grassland is sometimes called a **plagioclimax** (or sometimes referred to as a deflected climax). Other examples of plagioclimaxes are brought about by seasonal mowing, burning and weed spraying. Most of the land with which we are familiar comes into this category, from our lawns, playing fields and parks, which are mown to keep them as grassland, to grouse moors, where burning is carried out to keep the heather in the right condition for the game birds, and roadside verges, where until fairly recently both mowing and spraying to get rid of weeds has been carried out.

Sometimes a plagioclimax results from a combination of climatic conditions. When a hydrosere develops in an area of high rainfall and low temperatures, acid conditions prevail and the plant remains do not break down completely but accumulate, forming peat. The resulting climax community is moorland or bog instead of deciduous woodland, with large amounts of *Sphagnum*, a moss characteristic of wet, acid conditions.

Plant succession can be observed in sand dune systems (Figure 3.8), where it is possible to see all the seral stages of a primary succession. Nearest the water's edge are the newly formed **young dunes**, characterised by plants such as sea holly (*Eryngium*) and sea spurge (*Ammophila*), which are highly adapted to living in dry conditions. Behind the young dunes are **yellow dunes**, reaching heights of up to 20 m, mainly colonised by marram grass, but still showing large areas of bare sand. The colonisation is not complete and these dunes are often referred to as **partially fixed**. Behind these are **grey dunes**, where the sand is more stable and there is a greater diversity of plants. The ridges of dunes are separated from each other by **slacks**, which are low-lying damper areas with their own characteristic plants. Further back, grassland or heathland may become established. In some situations, a scrub of brambles (*Rubus*), elder (*Sambucus*), hawthorn (*Crataegus*) and dwarf willow (*Salix*) develops, which may eventually give rise to woodland. This is illustrated in Figure 3.9.

Figure 3.8 Sand dunes in North Wales showing Ammophila *colonising sand near the sea (in the foreground) and dunes building up and stabilising (in the background). The fences have been placed across sandy areas as a conservation measure to help maintain the dune system.*

Secondary succession in abandoned farmland

Agricultural crops in southeast England include annual plants such as wheat, barley and oil-seed rape. Traditionally, after the crops are harvested, fields are ploughed and prepared for the following year's crops. If such fields are left, without any interference, there is a progressive change in the structure of both plant and animal communities. Within a relatively short period of time, perhaps five years, the land becomes colonised by a mixture of herbaceous and shrubby, woody plants, collectively referred to as ruderals (plants that grow on wasteland or rubbish tips). These plants form a closed herb community and include plants such as nettles, thistles, willow herb, elder and brambles. Eventually, over a longer period of time, small trees become established and a climax community, such as mixed deciduous woodland, develops. As the plant communities become progressively more complex, that is, the numbers of different species increases; there is a corresponding increase in the complexity of the animal communities.

QUESTIONS

• Draw a food web to show the interactions between some of the species found in oak woodland.

• Find out whether there is a **special area of conservation** within the part of the country where you live, and the names of plants and animals which are conserved within this habitat.

[**Note:** Information about the EU Habitats Directive is available on the Internet]

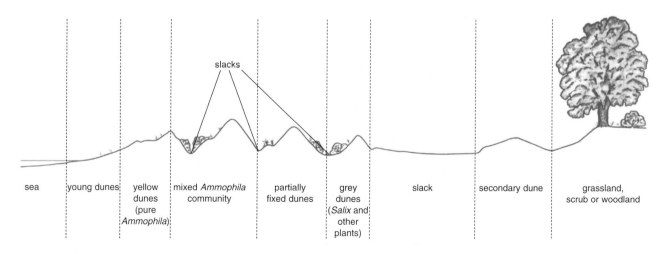

Figure 3.9 Succession in a sand dune system

ADDITIONAL MATERIAL

Zonation

Vertical zonation can be seen on mountains, where variations in the abiotic environment due to altitude have an effect on the spatial distribution of species. For each 100 m rise in altitude, the temperature drops by 0.5 °C and this has an effect on the plants and animals. Depending on the altitude, there is usually little soil on the top of a mountain so few plants, apart from lichens and some mosses, can grow. Lower down there is often grassland, scrub and eventually woodland at the bottom. In addition to the changes in temperature, there may be differences in rainfall, with one side of the mountain drier than the other. This type of zonation, illustrated in Figure 3.10, reflects the differences in the climatic conditions: a climax community develops within each zone.

Another example of zonation can be seen on a rocky shore, such as that depicted in Figure 3.11, where seaweeds and marine animal species occupy specific bands, or **zones**, between the low tide and high tide levels. The distribution of organisms

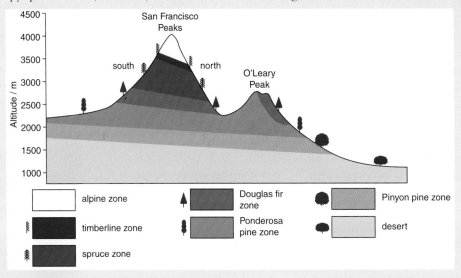

Figure 3.10 Zonation of vegetation on San Francisco Peaks, Arizona, as viewed from the south-east

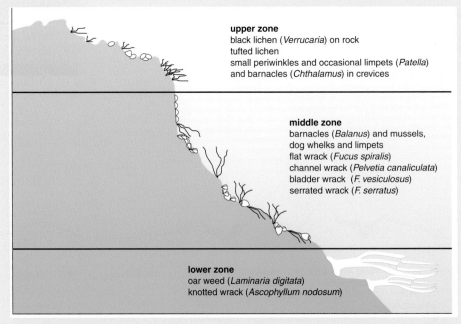

Figure 3.11 Zonation on a rocky shore

depends on their degree of tolerance to exposure and is affected by variations in the tidal height, which in turn cause variations in temperature and degree of desiccation. The tides range from the upper limit of **EHWS** (extreme high water of spring tides) to **ELWS** (extreme low water of spring tides). This range means that some organisms living high up on the shore will be covered with sea water only twice a month, whereas those living at the lower limit will be uncovered only twice a month. Between these limits, organisms will be exposed to the air for varying periods of time, most of them twice a day.

When the organisms are covered by the water, the temperature is more or less uniform, but when they are exposed to the air the temperature is much more variable. Organisms which can tolerate the greater variability of the air temperature tend to be found higher up the shore. Such organisms are also adapted to withstand drying out, many of them producing mucus coatings (for example, *Actinia* – sea anemones) or attaching themselves firmly to the rocks (for example, *Patella* – limpets).

The effects of wave action will depend on the slope of the shore and which way it is facing. Seaweed species growing on sheltered shores tend to be more abundant and larger than those on exposed shores, where wave action is greater. The following account of an investigation illustrates these differences.

A sample of ten specimens of the seaweed (*Fucus vesiculosus*) was collected at random from a sheltered shore, P, with little wave action, and a further sample of ten specimens was collected from a nearby exposed shore, Q, with considerable wave action. The length of the frond and diameter of the holdfast were measured for each specimen collected. The results of this investigation are given in Table 3.6.

Neither the zonation on the rocky shore nor that on the mountain show true succession in the same way as the sand dune system. The rocky shore and the mountain illustrate spatial distribution due to variations in the physical factors of the environment, whereas zonation in sand dunes enables us to observe a succession of seral stages which have taken place over a period of time.

Table 3.6 *Differences in length of seaweeds on exposed and sheltered shores*

Sample from sheltered shore P		Sample from exposed shore Q	
Length of frond / cm	Diameter of holdfast / cm	Length of frond / cm	Diameter of holdfast / cm
61.0	0.7	40.0	1.0
53.0	0.5	24.0	1.0
46.0	0.7	27.0	0.8
62.0	0.5	29.0	0.8
72.0	1.0	50.0	0.7
89.0	0.2	37.0	0.5
64.0	0.6	52.0	1.0
49.0	0.7	28.0	0.5
50.0	0.4	33.0	0.6
44.0	0.4	47.0	0.7

PRACTICAL   Determining water content of a soil sample

Materials

- Soil sample
- Evaporating dish or crucible
- Accurate balance
- Thermostatically controlled oven
- Desiccator
- Tongs

Method

1 Weigh the evaporating dish and record the mass.
2 Add about 10 g of soil to the dish and reweigh.
3 Subtract the mass of the dish to obtain the fresh mass of soil.
4 Place the dish plus soil in a thermostatically controlled oven at 105 °C. Leave for 24 hours.

5 Remove the dish and allow to cool in a desiccator. Record the mass.
6 Replace in the oven for 24 hours and repeat the previous stage until a constant mass is obtained.
7 Calculate the loss in mass of the soil sample. This represents the mass of water in the soil which can be expressed as a percentage of the fresh mass:

> Percentage of water in the soil
> = (loss in mass ÷ fresh mass) × 100

8 Use this method to compare the percentage of water in different soil samples.

PRACTICAL   Determining the pH of a soil sample

Materials

- Soil sample
- Test tube
- Barium sulphate
- Distilled water
- Universal indicator solution
- Colour chart

Method

1 Place about 1 cm of soil and 1 cm of barium sulphate in a test tube. Barium sulphate causes clay particles to settle leaving a clear solution.

2 Add 10 cm^3 of distilled water and shake the tube thoroughly.
3 Add a few drops of universal indicator solution and compare the colour with the colour chart against a white background.
4 Record the pH value.

This method has the advantage of being suitable for use in the field. More accurate results may be obtained by using a pH electrode and meter to find the pH of the soil solution.

PRACTICAL   Timed float method to determine current velocity

Materials

- Tape measure
- Stop watch
- Suitable float (an orange is recommended because it is conspicuous and will float mainly below the surface)

Method

1 Choose a suitable straight stretch of water. Measure the distance with the tape measure.
2 Accurately time the float over the measured distance. Repeat three times to obtain a mean.

3 Divide the mean time by the coefficient 0.85. This will give a more accurate velocity for the stream because the water at the surface flows faster than that beneath.
4 Calculate the velocity using the formula: velocity = distance ÷ time.

As an example:

Mean time of floats	= 14 sec
Mean time divided by 0.85	= 16.47 sec
Distance	= 10 m
Velocity	= 10 ÷ 16.47
	= 0.61 m sec^{-1}

| PRACTICAL | **A comparison of two closely related communities** |

Introduction

The aim of this investigation is to compare the distribution of species in two closely related communities and to consider the factors which may influence the distribution of organisms. The choice of sites to compare will, of course, depend upon access, but valuable data can be obtained using, for example, a footpath across a lawn or school field to investigate the effect of trampling. Other possible sites include:

- managed and unmanaged woodland, with the same dominant species
- north-facing and south-facing slopes
- grazed and ungrazed grassland
- football pitch and margins of the pitch
- well-drained and marshy land.

In each case, a **belt transect** is used to investigate changes in the communities. Ideally, the quadrats should be placed immediately after each other, although it may be more appropriate to sample at, for example, 1 metre intervals.

Materials

- Tape measure
- Marking pegs
- A quadrat frame, e.g. 0.5 m × 0.5 m
- Identification keys
- Recording sheets

Method

1 Set out the tape measure across the area you are going to study.
2 Place the quadrat at suitable intervals and, each time, record the presence of species, estimate their percentage cover, note the growth form of the plants and measure their maximum height.
3 If possible, obtain soil samples at each site to determine, for example, pH, moisture and humus content.

4 Repeat this procedure using a parallel transect.

Results and discussion

1 Draw a simple map of your study area and include relevant background information.
2 Record your results carefully and present using appropriate graphs.
3 Which abiotic factors show changes along your transect?
4 Consider the extent to which changes in plant communities can be explained by human activities and related biotic factors.

Further work

It is possible to make a quantitative comparison of contrasting communities by calculating a **diversity index**. In general, a complex community (consisting of a large number of different species but relatively few individuals of each species) is more stable than a community containing relatively few species. Polluted fresh water, for example, may contain very few species of invertebrates, but large populations of each.

A community can be evaluated by counting the numbers of individuals of each species. One way of expressing this relationship is **Simpson's Diversity Index (DI)**, which is based on the probability of randomly collecting a pair of organisms of the same species from a population. The higher the value of the calculated index, the greater the species diversity of the community. Simpson's Diversity Index is calculated using the formula:

$$DI = \frac{N(N-1)}{\Sigma n(n-1)}$$

where DI = diversity index, N = total number of individuals of all species and n = number of individuals of a species.

4

Populations and pest control

Populations and communities

A **population** is defined as a group of individuals of one species found in the same habitat. For example, a woodland habitat may include a population of bluebells (*Endymion non-scriptus*) and the leaf litter may contain a population of brown-lipped snails (*Cepaea nemoralis*) (see photograph Figure 8.3, page 110).

The term **community** is used to refer either to all the organisms, of all species, present in a particular habitat or to one group within the habitat, such as the plant community. The plant community in a woodland habitat will contain all species of trees, shrubs and herbaceous plants. The mollusc community in a pond might contain *Lymnaea stagnalis*, *L. peregra* and *Planorbis crista*.

Population size refers to the number of individuals of a species present in an area at a particular time. Populations in ecosystems interact and, as a result, the **population density** (that is, the number of individuals per unit area or per unit volume) changes with time.

Factors affecting population size

One of the fundamental characteristics of living organisms is that they reproduce and so populations tend to increase in number. A single bacterial cell, for example, may divide into two cells every 20 minutes under favourable conditions. This means that the number of cells in a culture doubles every 20 minutes. A graph of the number of cells plotted against time (referred to as a growth curve) shows **exponential growth** (Figure 4.1).

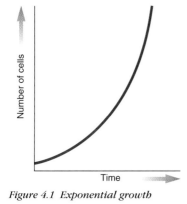

Figure 4.1 Exponential growth

This growth cannot continue indefinitely – the food supply may be limited, or the organisms may produce toxic by-products of metabolism. Within a relatively short period of time, the growth rate decreases and the cells stop dividing. This is referred to as the **stationary phase.** Eventually, cells may begin to die, so that the number of viable cells decreases. The culture has then entered the **death** (or **decline**) **phase** (Figure 4.2).

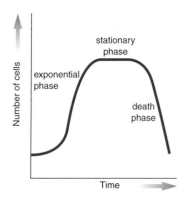

Figure 4.2 Growth curve for a bacterial population

Similar changes can occur in natural populations. For example, rapid growth of algal cells can occur in ponds or lakes as a result of enrichment with mineral ions such as nitrate. Figure 4.3 shows the growth and subsequent decline in numbers of the brown alga *Dinobryon divergens*.

Figure 4.4 shows another example of a change in a population, that of the numbers of sheep in Tasmania following their introduction to the island in 1814. The fine line shows the year-to-year variations in numbers; the thick line shows the overall population trend. Notice that the numbers increase up to a maximum and then the population levels out and remains approximately constant.

QUESTION

In Figure 4.3, what are the reasons for the decrease in numbers of algal cells between June and August?

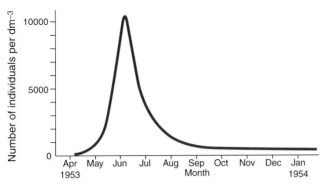

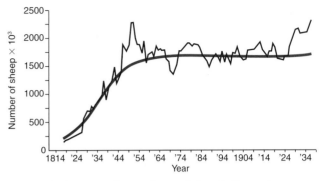

Figure 4.3 Changes in the population of the alga Dinobryon divergens *between April 1953 and January 1954*

Figure 4.4 Changes in the population of sheep in Tasmania following their introduction in 1814

Carrying capacity and environmental resistance

Although this population of sheep (Figure 4.4) could, theoretically, continue to increase, it does not and levels out at a population of about 1.5 million. The maximum size of a population that a particular environment can support is called that environment's **carrying capacity.** Various environmental factors, referred to as **environmental resistance**, reduce the growth rate of a population. These factors include disease, competition, predation and unfavourable climatic conditions.

The number of individuals in a population changes as a result of fluctuations in the birth and death rates. To illustrate this, we will consider the results of a study on the numbers of great tits (*Parus major*) in Sweden. Figure 4.5 shows the changes in both the population of great tits and the availability of beech seeds upon which they feed. Dramatic increases in the numbers of these birds followed large crops of beech seeds. This extra available food meant that more juvenile birds were able to survive the winter and their numbers increased the size of the population the following spring.

Populations do not keep increasing indefinitely. Population size may be limited by either increases in mortality (deaths) or by decreases in natality (births). Mortality and natality are often expressed as rates, for example, a mortality rate of 2/100 (or 2 per cent) represents two deaths in a population of 100.

Density-dependent and density-independent factors

Death rates that change with population density are described as density-dependent mortality rates. Density-dependent mortality is an important factor in limiting the growth of a population when it reaches a high density.

Density-dependent factors are biotic factors which are influenced by the density of the population and include disease and predation. If the population density increases then the chance of a disease being transmitted from individual to individual increases. **Density-independent** factors are not influenced by the density of the population and are typically abiotic, such as extremes of temperature or acid rain.

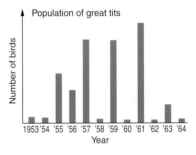

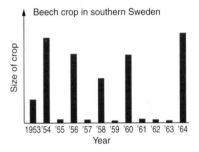

*Figure 4.5 Changes in the breeding numbers of great tits (*Parus major*) and the size of the beech crop in southern Sweden*

> ### DEFINITION
>
> **Density-dependent** factors are biotic factors, such as competition, predation and disease, which increases as the population increases. **Density-independent** factors are abiotic factors, such as temperature, light intensity and humidity, which have an effect on populations, but which are unrelated to the population size.

POPULATIONS AND PEST CONTROL

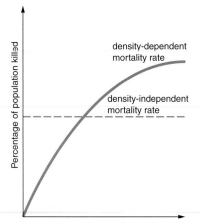

Figure 4.6 Density-dependent and density-independent mortality rates compared

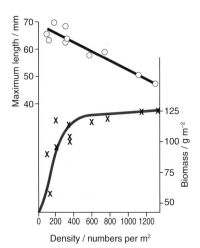

Figure 4.7 Effect of intraspecific competition on length (○) and biomass (X) of the limpet Patella cochlear

Results of a detailed study of the population of tawny owls in Wytham Wood, near Oxford, showed that the mortality rate of young owls was much greater in years when the wood already had a high owl population. In other years, when the overall population was low, many more young owls survived and established territories of their own. The mortality rate of young owls is therefore density dependent and this has an important regulating effect on the population density. Figure 4.6 compares density-dependent and density-independent mortality rates. The solid line shows an example of a density-dependent mortality rate, where the mortality rate is proportional to the population density. The dotted line indicates a density-independent mortality rate, showing that a fixed percentage of the population is likely to die irrespective of the population density. In the case of tawny owls, this could be a result of some of the birds being killed by the winter cold.

Density-dependent mortality factors are always biotic (such as competition for food or space), whereas density-independent mortality factors may be either abiotic (such as winter cold) or biotic.

Intraspecific and interspecific competition

In an ecosystem, organisms which are using a limited resource, such as food, space or light, will compete with each other for that resource. As a result of this competition, some individuals may have a reduced growth rate, or increased risk of mortality. Competition therefore reduces the rate of population growth.

Intraspecific competition

This occurs between organisms of the *same* species. These are likely to have similar resource requirements, so they compete for any that are in limited supply. The more competitors there are, the greater the effect of intraspecific competition is likely to be. In other words, intraspecific competition is density dependent. Competition between tawny owls, described previously, is one example of intraspecific competition. As another example, we will consider the effect of intraspecific competition on the limpet *Patella cochlear*. This is a marine mollusc which feeds by grazing on algae which grow on the limpets' rocky habitat. As the density of limpets increases, competition for food also increases. The effects of population density on the maximum length and biomass of *Patella cochlear* are shown in Figure 4.7. As the population density rises, intraspecific competition increases, resulting in a reduction in length but maintaining an approximately constant population biomass.

Interspecific competition

In a complex ecosystem, with many different populations competing for resources such as space, food or light, it is inevitable that competition will occur between *different* species. This is known as **interspecific competition.**

Over 50 years ago, the Russian ecologist G.F. Gause proposed the **principle of competitive exclusion.** This principle states that, if two species are competing with each other for the same limited resource, then one species will be able to use that resource more efficiently than the other, so the former species will eventually eliminate the latter. Gause carried out a series of experiments using two species of *Paramecium: P. caudatum* and *P. aurelia* (kingdom Protoctista,

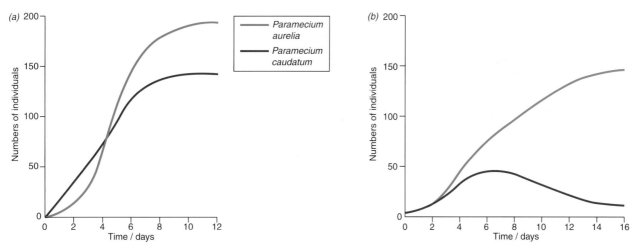

Figure 4.8 Interspecific competition between two species of Paramecium *(a) cultured separately; (b) cultured together*

phylum Ciliophora). These experiments showed that, when these two species are grown together in a mixed culture, *P. aurelia* drives *P. caudatum* to extinction; in other words, *P. caudatum* had been competitively excluded. This is because *P. aurelia* uses the available food resources more efficiently, and reproduces more quickly than *P. caudatum* under these conditions (Figure 4.8).

Predator–prey relationships

A **predator** is an organism which kills and feeds on all or parts of animals, referred to as the **prey**. Predation is an important biotic factor which influences the abundance of organisms in an ecosystem.

The size of a population of a predator will depend on the availability of its food source, so the larger the prey population, the larger the population of predators it can support. However, increased numbers of predators will decrease the numbers of prey organisms. The populations of each would be expected to show a series of regular cycles, where the population of predators is out of phase with the population of prey. The prey population peaks before that of the predator. This is illustrated in Figure 4.9, which shows a hypothetical model of predator–prey numbers.

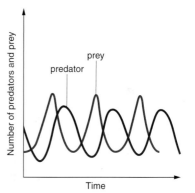

Figure 4.9 Model of predator–prey interaction

This model has been supported by experimental work involving two species of mites. *Typhlodromus occidentalis*, a predatory mite, was added to cultures of its prey, *Eotetranychus sexmaculatus*, a herbivorous mite, feeding on oranges in a tray. Figure 4.10 shows the resulting fluctuations in the populations of each species, where the numbers of each can be seen to oscillate and the cycles are slightly out of phase with each other.

Similar fluctuations in populations can be seen in, for example, the numbers of glasshouse whiteflies (*Trialeurodes vaporariorum*) and the predatory insect *Encarsia formosa*. This predator is used as a means of biological control of the whitefly, which is discussed on page 58.

There is evidence that predator–prey interactions are responsible for the fluctuating numbers of snowshoe hares (*Lepus americanus*) in northern

QUESTION

Explain why the peaks in the numbers of prey organisms occur before the peaks in the number of predators.

51

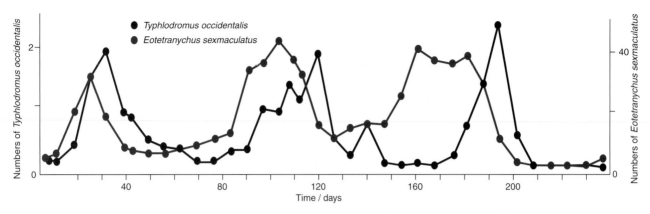

Figure 4.10 Predator–prey interaction between two species of mite

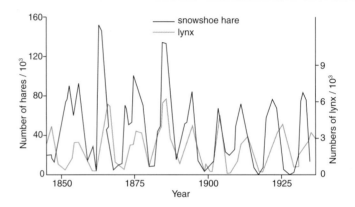

Figure 4.11 Predator–prey interaction between lynx and snowshoe hare

Figure 4.12 North American lupin aphids (Macrosyphum albifrons) feeding on lupin stem. (The large female on the left is giving birth to a live nymph.) Aphids cause direct damage by piercing the phloem and sucking out sap. They also cause indirect damage by transmitting viruses which cause disease.

Canada. One of its main predators is the lynx (*Lynx canadensis*), which shows similar, but out of phase, fluctuations in population density (Figure 4.11).

Control of insect populations

Insects are are usually present in considerable numbers in any habitat, such as grassland, roadside verges, hedges, woodland and cultivated land. They occur in the soil, among decaying litter on the ground and feeding on plants or other animals in the area. Estimates suggest there can be as many as 25 million insects in a hectare of soil, and perhaps 25 000 in flight over a hectare of land at any one time. A sugar beet crop infested with black bean aphids could even carry up to 200 million aphids per hectare.

As with any populations of organisms, numbers of insects within a population are never static and, in any defined place, are liable to fluctuate over a period of time. Actual numbers in a population depend on the number of individuals of the species that are being born or dying, and those migrating into or out of the area. Changes in population numbers are influenced by season, weather conditions and competition between organisms, as described earlier in this chapter. Remember also that insects are part of food webs so that any alteration in numbers of one species is likely to affect other interrelated organisms.

Insects are described as **pests** when they cause damage or nuisance, directly or indirectly, to crop plants, domesticated animals or humans (Figure 4.12). The discussion here will focus on pests in relation to crop plants in which there may be losses in yield or quality of the produce as a result of insect pests. Considerable expenditure is devoted to the control of insect pests of agricultural and horticultural crops. If there were no insect pests, on a global scale food production could probably be increased by 30 per cent. A successful insect pest is often one that can establish itself and spread rapidly in the crop. A high rate of reproduction is a feature that may contribute to success. The monocultures associated with modern intensive agricultural systems often present an ideal situation for the insect pest by providing easy access to an abundant food supply, with few if any natural predators in the area.

Chemical control of insect pests

An **insecticide** is a chemical substance that can kill or repel insects. Naturally occurring insecticidal chemicals, which have been used for hundreds of years, include pyrethrum from the flower of *Pyrethrum cinereafolium* (in the daisy family) and nicotine from *Nicotiana tabacum* (the tobacco plant). Many artificial chemical substances have now been synthesised for use as insecticides. These insecticides are complex organic substances and fall into four main groups: the **organochlorines**, **organophosphates**, **carbamates** and **pyrethroids** (see Extension Material and Table 4.1). You will see, for example, that the full name of DDT (dichlorodiphenyltrichloroethane) is 1,1,1-trichloro-2,2-bis-(4-chorophenyl)ethane.

Insecticides get into the insects in different ways. Some insecticides have their effect by direct **contact** with the insect and penetrate the waxy cuticle on the surface of the insect. **Fumigants** (see Extension Material) are inhaled by the insects. Others act as **stomach poisons** and are taken in when the insect eats the leaf. **Systemic** insecticides are taken into the plant by absorption through the leaves or roots, or by the seed when it germinates, then circulate through the plant. Systemic insecticides then enter the insects when they feed on the sap of the plant and are particularly useful for controlling sap-sucking insects such as aphids.

In most cases, the toxicity of an insecticide is due to its interference with the nervous or respiratory systems. Often an insecticide is also toxic to other species, including harmless insects, humans or other mammals. Some of these species may even be beneficial due to their predatory action on the pests (hoverflies and ladybirds on aphids, for example) or because they carry out pollination (bees). Systemic insecticides have an advantage in that they only affect insects that feed on the plant. They are thus useful for the control of aphids sucking sap but do not affect predators or other harmless insects walking over the leaves. Pirimicarb is an example of an insecticide (systemic and fumigant) that has a biochemical selectivity for aphids and some flies, but a very low toxicity to mammals. When applying insecticides, strict precautions must be taken to avoid unwanted harmful effects on humans and other organisms.

BACKGROUND

The Colorado beetle (*Leptinotarsa decemlineata*) illustrates how friend can turn into foe. In 1824 the Colorado beetle was described by a collector from the Rocky Mountains in north America. This attractive, striped beetle was considered a rarity. It fed on the weed buffalo-bur, a member of the potato family. Thirty years later, settlers to the region brought potatoes as one of their crops. The potatoes provided a nutritious and plentiful new food source for the beetles and their effect on the crops was devastating. Populations of the Colorado beetle rapidly increased and spread eastwards across the USA in the second half of the 19th century, causing famine as the crops failed to produce enough food. Finally a decision was taken to spray the potato crops with Paris Green, an insecticide containing arsenic, as a means of controlling the beetle. The Colorado beetle is now treated as a pest on an international scale, with strict regulations governing procedures to be taken if any are found.

EXTENSION MATERIAL

More information about chemical insecticides

Table 4.1 *The main groups of chemical insecticides*

Insecticide group	Examples	Mode of action	Behaviour in plant, relative toxicity and other comments
organochlorines e.g. DDT Cl_3C—CH (with two chlorophenyl rings, Cl)	DDT, HCH, aldrin, dieldrin, endosulfan	DDT inhibits enzyme cytochrome oxidase, destabilises nervous system by interfering with permeability of nerve axon membrane	• low toxicity to humans, so relatively safe when being applied • slow breakdown, so persistent in soil • broad spectrum **contact** insecticide
organophosphates e.g. parathion C_2H_5O, C_2H_5O, P=S, O— ring —NO_2	parathion, malathion, dimethoate, metasystox, schradan	combine with cholinesterase, thus inhibiting hydrolysis of acetylcholine produced at nerve endings, so interfere with transmission of nerve impulses across synapses	• high toxicity to humans • easily broken down so less persistent in soil • considerable flexibility in the group, most show **systemic** action; malathion is a **fumigant** insecticide
carbamates e.g. carbaryl (naphthalene ring) O—CO—NH—CH_3	carbaryl, carbofuran, aldicarb, methomyl, pirimicarb	interfere with the nervous system by acting as competitors with cholinesterase, thus inhibiting hydrolysis of acetylcholine	• some very toxic to humans (carbofuran, aldicarb) • persistence lies between organochlorines and organophosphates • pirimicarb has low mammalian toxicity, biochemical selectivity for aphids and some flies • methomyl has good **contact** action, is also a **fumigant** and to some extent has **systemic** action
pyrethroids e.g. resmethrin MeCCH, Me, Me —COCH$_2$, O, —CH$_2$Ph	(*natural*) pyrethrum, nicotine, rotenone (*synthetic*) allethrin, cypermethrin, resmethrin	similar to DDT	• high toxicity to insects, low toxicity to mammals • similar mechanism to DDT, so some cross-resistance • **contact** insecticide, so damaging to natural populations of insects

Insecticides are usually dissolved in water or oil and applied by spraying in fine droplets. In some cases they may be dispersed on an inert solid carrier. In closed spaces, such as glasshouses, the insecticide may be burned to give off smoke (**fumigant** insecticides). Spraying may be done from small hand-operated tanks or by means of large-scale machinery using tractors or aircraft. It is important that the sprayed insecticide reaches the target (either the insect itself or the plant) at a suitable time and in the required concentration, without drifting or causing damage elsewhere, and that the insecticide remains on or in its target long enough to exert its effect on the insect pest. In practice, often only a small fraction of the insecticide actually lands on the required target.

Over a period of time, the effectiveness of some insecticides has become less because of the development of resistance within the insect populations. The situation provides an example of genetic change in a population in response to selection pressure. Within the gene pool of the population, there are likely to be some individuals that are not killed by the insecticide, or have a natural resistance to its harmful effects. If these insects survive and reproduce, their offspring may inherit the characters which produced the resistance. In

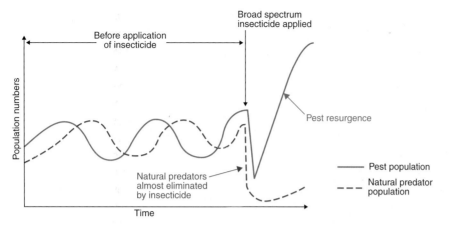

Figure 4.13 Resurgence of a pest after application of a broad-spectrum insecticide. Before application of this insecticide, the pest population numbers in a given area (density) are controlled by natural predators – numbers of the pest increase, then the natural predators increase after a short time lag. After application of the insecticide, the natural predators are also killed so they can no longer control the numbers of the pest. Surviving pests may then show a population explosion.

subsequent generations an increasing proportion of the population is likely to show resistance to the insecticide.

Another problem associated with the use of insecticides is that of **resurgence**, illustrated in Figure 4.13. When no pesticide is applied, population numbers of a pest and its natural predators are likely to fluctuate around an equilibrium level. However, if a broad-spectrum pesticide is used, natural predators are likely to be killed as well as the pest. With lower numbers of predators, any surviving pests may increase very rapidly leading to a population explosion.

Bioaccumulation of non-biodegradable toxins

Insecticides differ with respect to the length of time they persist in the plant, or in the soil, before breaking down. The term **biodegradable** is used to describe substances that can be broken down by the action of living organisms (often bacteria) and with the implication that the residues are non-toxic; **non-biodegradable** indicates that the substance cannot be broken down by living organisms or that the breakdown is very slow. The organochlorines (see Extension Material) are relatively stable and resistant to breakdown, whereas most organophosphates are biodegradable and rapidly broken down by metabolic reactions in animals and excreted as harmless substances.

The persistence associated with organochlorines may at first sight appear to be an advantage in terms of controlling the pest insect, but, over a period of time, can result in undesirable toxic side-effects. This is well illustrated by **DDT**. In the 1940s and 1950s, DDT was welcomed as a solution to many pest problems, for control of pests on crops and to minimise the spread of malaria by killing the mosquito (a vector of the pathogen which causes malaria). However, since the 1960s, there has been increasing concern over the longer term effects of using DDT and other persistent insecticides.

Persistent insecticides, such as DDT, do not break down into simple substances that can be excreted from the insect, or they take a very long time to break down. The insecticide molecule remains in the body of the insect so, when it is eaten, the insecticide (or a derivative of it) passes into the next organism, and so on along the food chain. The insecticide molecule may still be harmful to these other organisms. In addition, toxic residues from the insecticides in the soil may seep into watercourses and contaminate them, or the insecticides may persist on crop plants and then be found in human food. In this way, the harmful effects of the insecticide can spread through food chains and food webs to a wider range of organisms. In the 1960s, residues of organochlorines were detected in a very wide range of organisms, including humans and many food sources. This led to considerable concern over the widespread use of insecticides.

There is also evidence that persistent insecticides accumulate at progressively higher concentrations as they pass along the food chain. That is, the concentration of insecticides that remains in organisms of the food web tends to increase at higher trophic levels, compared with that found in organisms at lower trophic levels. The accumulation is greatest in the top carnivores. This is known as **bioaccumulation**.

Convincing evidence for bioaccumulation has been obtained from studies of British birds and their eggs. Some data from these studies are presented in Figure 4.14. These focus on persistent organochlorines, particularly **DDT** and **aldrin** and their derivatives (DDE and dieldrin respectively, both of which are more persistent than the original pesticide). The data in Figure 4.14 show the relative concentrations of organochlorine residues in breast muscle of the birds and of DDE and dieldrin in egg shells. Look first at the type of diet to determine whether the organism is a herbivore, omnivore or carnivore. This also tells you its position in a food chain or food web. These data clearly illustrate the extent to which these pesticides were found to have accumulated in the higher trophic levels, particularly in the top carnivores.

We can explain the process of bioaccumulation by reference to a simple food chain. (In this food chain, bird A is a secondary consumer and bird B is a top carnivore.)

crop plant → insect → bird A → bird B

The insect pest eats the crop plant and takes in the insecticide (either through the plant or by direct contact with the chemical), but the insecticide molecule does not break down inside the insect. Bird A eats a large number of these insects, thus gathering a relatively high concentration of the insecticide inside its body. Similarly because bird B eats a large number of bird A, an even higher concentration of insecticide accumulates inside bird B.

Figure 4.14 Persistent pesticides and how they pass along a food chain: (a) residues of organochlorine insecticides in breast muscle of some British birds; (b) dieldrin and DDE found in eggs. Concentrations are given in parts per million (ppm) by mass. The data refer to studies carried out in the 1960s and illustrate that pesticides were found to have accumulated in the higher trophic levels.

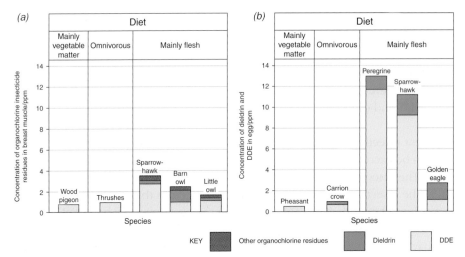

Biological control of insect pests

Biological control exploits interactions between organisms. The term usually implies the deliberate introduction of one species to control another species (the pest), either through a predator–prey relationship or as a parasite that weakens or causes disease in the pest. The controlling organism may already be present in the area as a natural enemy (but probably in low numbers) or may be an alien, introduced to the area by humans.

The biological control agent may be another insect, or control may be through viruses, bacteria or fungi. Some examples of biological control agents used against insect pests are given in Table 4.2. The success of a control agent depends on the nature of the relationship between the two species and on the population dynamics. If the control organism is introduced in low numbers, it may take time to build up sufficient numbers to harm the pest; meanwhile the pest may have done considerable damage to the crop. Generally the aim is to keep the damage from the pest at a level acceptable to the grower rather than to eliminate the pest completely (Figure 4.15). If the control organism is present in large numbers or is very effective at eliminating the pest, the control organism is then deprived of its food source. Numbers of both the pest and the control organism are likely to fluctuate, since one depends upon the other. In some cases the biological control organisms are effective only in a confined space and where conditions can be controlled, such as inside a glasshouse.

An understanding of the ecology and behaviour of both organisms is needed to ensure the control organism is introduced at appropriate times and in suitable numbers to exert effective control.

Some examples will help to illustrate this.

- *Phytoseiulus persimilis* is a predatory mite, which feeds only on the red spider mite (*Panonychus ulmi*), and can be used successfully as a biological control agent. Red spider mites are pests of a number of glasshouse crops, including cucumbers and tomatoes. In Britain, the main attack by red spider mite starts when overwintering females emerge from the soil and walls in April or May. Even though the predatory mite can survive at lower temperatures (down to 5 °C) it would be inappropriate and uneconomic to

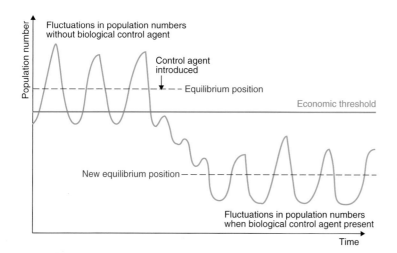

Figure 4.15 Biological control, showing how the numbers of a pest insect in a given area (density) can be reduced from a high equilibrium level to a new lower equilibrium level. The aim is to bring the numbers of the insect pest below the level which causes unacceptable economic damage.

EXTENSION MATERIAL

Table 4.2 *Examples of biological control of insect pests*

Biological control agent		Insect pest	Example of use and other comments
viruses	baculoviruses	larval stages of butterflies and moths; ants, bees and wasps; flies, gnats, midges, beetles, caddis flies	• different strains of virus attack different species of insect • used in USA to protect cotton and fir trees, in France for vegetables, in Brazil for soya bean
bacteria	*Bacillus thuringiensis*	larval stages of butterflies and moths; beetles, flies	• different subspecies produce toxins active against different insects • used widely to protect edible and ornamental flowering plants
fungi	*Verticillium lecanii*	aphids and whitefly	• used on cucumber, eggplant, chrysanthemums
nematodes	*Heterorhabditis megidis*	vine weevil larvae	• used to protect bedding plants, e.g. primulas
insects	*Encarsia formosa*	aphids and whitefly	• parasitic wasp, used widely for control of whitefly in glasshouses
	Phytoseiulus persimilis	mites	• predatory mite, used for control of red spider mite in glasshouses

introduce the predatory mite before the red spider mite has emerged.

- Whitefly is another pest of glasshouse crops. Control of whitefly by the parasitic wasp *Encarsia formosa* is more critical in terms of temperature requirements. The wasp does not overwinter in Britain in unheated glasshouses because it does not survive below 10 °C, so fresh supplies must be introduced at an appropriate time. The wasp requires a temperature above about 18 °C to keep pace with increases in numbers of whitefly. However, at temperatures above 26 °C the wasp reproduces much more rapidly than the whitefly and is likely to eliminate the pest. This means the wasp population also dies out and needs to be replenished if further control of the pest is required.

Encouraging natural predators

Natural predators of insect pests are likely to be present in hedges or uncultivated margins around the edges of fields. These natural predators can be encouraged by maintaining suitable habitats close to or amongst the crops. One way is to create strips of uncultivated land within fields. The strips (sometimes known as **beetle banks**) are sown with suitable vegetation, which allows natural predators to overwinter close to the site of the crop. The spacing of the strips is such that predators (such as hoverflies and lacewings) can penetrate into most of the crop when foraging for food. Even though this practice means a small reduction in land available for growing the crop, this is more than compensated for by the reduction in pest damage. Such habitat areas give further benefit by creating diversity within stretches of land dominated by monoculture.

Intercropping provides an opportunity for predators to live on one crop and attack the prey on the adjacent crop (Figure 4.16). As examples, undersowing with ryegrass provides suitable conditions for ladybirds as a means of controlling aphids on cereals; cabbage intercropped with red and white clover helps to regulate cabbage aphids, probably because of increase in ground beetles. Interruption of one crop by another may also reduce the spread of insect pests through the affected crop. Similarly, mulches and even weeds can be beneficial by providing a suitable habitat for natural predators. Other cultivation techniques can be used to attract or maintain populations of natural predators (Figure 4.16).

Comparing control by chemical and biological methods

A review of the ways used to control insect pests on crops shows that each approach has both advantages and disadvantages. Some of the features are summarised below.

Use of toxic chemicals (insecticides) to kill or repel pests:
- Most insecticides are general rather than specific in their effects.
- They lose effectiveness when resistant strains of pest insect increase in the population.
- There is concern over the effects of toxicity on other animal species which might naturally help to control pests.
- Persistence of some insecticides means their effects last over a period of time and spread through food webs to other organisms.
- Some insecticides taint or spoil the flavour of food crops.
- High costs are involved (research development, manufacture, machinery to apply the insecticide).

Deliberate biological control by introducing or increasing the numbers of predators or parasites of the pest insect:
- With biological control, a specific pest species or group of pests can be targeted.
- Biological control acts relatively slowly and may be unpredictable.
- Usually the pest is not eliminated completely.
- The control population can be self-perpetuating (but not if the control species completely eliminates the pest).
- Careful control is required with regard to timing of the release.
- Generally there are no side-effects.
- Biological control imposes limits on the use of pesticides to control other pests on the same crop, which may not be controlled by the biological agent.
- Biological control is relatively inexpensive, both in development costs and application.

Relying on natural predators:
- The pest population is kept at a low level rather than being eliminated.
- The level of control is unpredictable; populations are liable to fluctuate.
- These methods are natural, and do not damage the environment.
- Costs are minimal.
- Natural predators can be encouraged by maintenance of suitable habitats in the vicinity of the crops (hedges, beetle banks, intercropping, rotation of crops).

Figure 4.16 (top) Intercropping – wheat intercropped with maize, in western China; (bottom) blue flowers of Phacelia tanacetifolia *can be used to attract natural predators. Strips of* Phacelia *grown along the margins of cereal crops attract hoverflies, which lay their eggs within the crop. Hoverfly larvae are voracious predators of aphids.*

Integrated pest management (IPM)

World-wide, there are pressures (on farmers and other growers) to increase crop yields. At the same time, there is increasing awareness of the disadvantages of control measures that are dependent entirely on the use of pesticides. With better understanding of the ecology of natural ecosystems and the factors affecting crop growth, together with the relationship of specific pests with the crops, a combination of approaches is often the best way of achieving control of populations of insect pests.

Development and implementation of such strategies are described as **integrated pest management (IPM)**. IPM attempts to exploit control of pests by natural enemies (including predators and parasites of the pest) or by the use of artificially introduced biological control agents, and to utilise chemical pesticides only when necessary. It requires a coordinated or managed approach, using knowledge of the ecosystem and information relating to prevailing conditions, including weather or the prediction of an imminent attack by particular pests. Often the systems used to gather the relevant information are highly sophisticated so that an attack can be predicted with considerable precision.

We can consider two examples to illustrate the principles behind IPM. The first example highlights a management strategy that exploits the activities of natural predators and integrates this with some limited use of insecticides. The second example describes an integrated approach involving natural enemies, biological control agents and insecticides.

First we look at growing rice paddy in tropical regions of south-east Asia. The rice plant is subject to attack by a mass of insect pests and, in the absence of attempts at control of the pests, crop losses can be considerable. With two or three rice crops in a year, following on or even overlapping with each other, this monoculture provides an ideal breeding ground for insect pests. There are, however, a rich number of natural predators and parasites that interact in a complex way with the various pests that are associated with the rice plants. These natural enemies can have a significant role in controlling the pests. Unfortunately, the effect of these natural enemies can be severely diminished by application of chemical insecticides, particularly when used early in the season. The insecticides tend to eliminate the natural predators and thus destroy the ecological balance. This then allows the number of the insect pests to increase, without the benefit of interaction with and control by the natural predators, resulting in damage to the rice plants and followed by loss in yield of the crop. If, however, insecticides are not used early in the season, the natural predators are able to exert some control over invading pests. At a later stage in the growing cycle, it may be advisable to make limited use of an insecticide, to help control peaks in numbers of pests. It is now recognised that early applications of insecticides are often unnecessary (and costly) and actually counterproductive because of their effect of destroying the natural control systems.

The second example illustrates how an integrated approach helped overcome a specific problem. The spotted alfalfa aphid was first seen in California in 1954.

Organophosphates were used to control the aphids but by the end of the 1950s resistance to the insecticide had developed and losses of crops (such as lucerne) had become critical. The usual response would be to increase the application of pesticide, but instead a *reduced* dosage of insecticide (dimethoate) was applied. Even though this meant fewer aphids were killed by insecticide, it allowed some of the natural enemies to survive and these helped to control the aphids. Two further measures were then taken which helped to control the pest. Firstly, the local natural enemies were reinforced by introducing additional parasitic species. Secondly, the lucerne crop was harvested in strips. This allowed some aphids to remain on the newly cut strips with their natural enemies, while at the same time older strips of lucerne were treated by insecticides.

This sort of integrated approach does require an understanding of the sequence of events in terms of its ecology, and management so that appropriate strategies are implemented and at the right time. These examples help to explain the term **integrated pest management** and give an appreciation that the precise strategy depends on the particular situation, including the prevailing patterns of weather and pests involved. It is encouraging that, for example, support groups are in existence within farming communities in some rice growing areas. This support network attempts to educate farmers and share expertise so that they understand the principles involved and, in their local situation, how best to exploit natural pests and minimise the use of insecticides. Ideally, the farmer becomes the expert in knowing what to do, when to do it and why. Integrated pest management really tries to encourage farmers or other growers to work through an understanding of the ecosystem and to manage their pest control strategy rather than just applying insecticides as the only means of controlling insect pests.

QUESTIONS

- How does knowledge of the biology and ecology of a pest help in deciding how to manage the control methods used? Think about timing of application (of insecticide or of biological control agent).
- What cultivation techniques would you suggest to someone who did not wish to use chemical insecticides?
- What features would you look for if you were asked to develop the 'perfect pesticide'?

POPULATIONS AND PEST CONTROL

Estimating population size using the Lincoln Index

Introduction

The **capture-mark-recapture** method is a technique for estimating the size of a population of organisms. A sample of the population is taken and these organisms are marked in some way so that they can be identified later. They are then released and allowed to disperse into the population. A second sample is then taken and the numbers of marked organisms recaptured, and those captured that are unmarked, are recorded. A formula, referred to as the **Lincoln Index**, is then used to estimate the total population size. There are a number of important assumptions in this method:

- the mark has no effect on the organisms
- the mark persists during the investigation
- the marked organisms disperse randomly throughout the whole population
- the population is closed, that is, no migration of organisms
- no births or deaths occur during the investigation.

The formula for the Lincoln Index is shown below:

$$N = \frac{S_1 \times S_2}{R}$$

where N = the estimated total population size,
S_1 = the number of organisms marked and released,
S_2 = the number of organisms captured in the second sample and
R = the number of marked organisms recaptured.

Full details of this method, and suggestions for its use in field work, are included in *Tools, Techniques and Assessment in Biology*. The aim of this practical is to use a laboratory model to illustrate the principle of the capture-mark-recapture method. Larvae of the beetle *Tenebrio molitor*, known as mealworms, are suggested for this investigation. Mealworms are usually available at pet shops, or may be obtained from Blades Biological or Philip Harris Education.

Materials

- Culture of mealworms in bran (this should contain at least 100 larvae)
- Plastic weighing boat, or similar, to remove samples
- Permanent marker pen, such as an overhead projector pen

Method

1 Take a sample of mealworms from the population. If there are fewer than 20 mealworms in this sample, take a larger sample.
2 Count and record the number of larvae in the sample (S_1), and mark each one with a coloured dot.
3 Return these larvae to the culture and leave for a standard time to allow the larvae to mix.
4 Take a second sample, note and record the total number of larvae present in this sample (S_2) and the number of these which are marked (R).

Results and discussion

1 Use the Lincoln Index formula to calculate the estimated population size.
2 Count the actual number of mealworms present in the population.
3 Compare your calculated value with the actual value and suggest reasons for any differences.
4 Consider how far this laboratory exercise reflects a field work investigation, such as estimating the size of a population of snails in grassland.

Further work

1 You could simulate the mark-release-recapture method using a beaker containing coloured beads, or dried peas. They must be thoroughly shaken to mix before taking your second sample!
2 Combine the capture-mark-recapture method with another method for determining population density. Take a sample as in step 1 of the method above, and count the number of mealworms present. Find the volume of this sample, for example, by using a measuring cylinder. Now find the total volume of the culture and calculate the estimated number of mealworms present. Compare your results using the two methods.
3 Use the capture-mark-recapture method to estimate numbers of ground beetles caught in pitfall traps, woodlice under logs, or frog hoppers in grassland. Frog hoppers are insects found, for example, in rough grassland. They are collected using a sweep net and pooter, then marked with a small dot of quick-drying paint or coloured nail varnish applied with the point of a pin, on their forewings. When the paint has dried, release the frog hoppers by scattering them over the area, then take a second sample, using a sweep net, at least 3 to 4 hours later.

Conservation

Conservation and biodiversity

Conservation has been defined as the protection and preservation of natural resources and of the environment. Broadly, when applied to ecosystems, the aims of conservation are to:

- conserve a range of different habitats so that they are not lost or drastically reduced through agriculture, urbanisation or other pressures of human populations (see *Exchange and Transport, Energy and Ecosystems*, Chapter 9)
- encourage biodiversity within a range of habitats.

On a global scale, maintenance of species diversity depends on there being a diversity of habitats and the communities associated with those habitats. The same principles can be applied to reduce the chance of rare species becoming extinct. Conservation of habitats and the communities within them can be achieved in specially designated nature reserves but can also be integrated with the farming systems of the region or encouraged in an urban situation. Conservation can also be effected on quite a small scale in, for example, a churchyard or a garden.

Conservation is dynamic rather than static and is achieved through active intervention and management rather than passive preservation. In this chapter we look at a number of specific examples to see how conservation is achieved. In particular, emphasis is given to the management strategies which can be adopted to encourage biodiversity in those particular habitats. [**Note:** *While these examples are not mandatory as part of the Edexcel specification, they should contribute to your understanding of this topic and enable you to interpret other situations which you may meet, in your studies or in the assessment tests.*]

Figure 5.1 Example of a succession – development from grassland to woodland: (1) mown frequently in growing season; (2) mown once or twice a year; (3) grassy area not mown; (4) woodland established

Succession and species diversity

Any community of plants and animals is strongly influenced by the physical factors of the surrounding environment. Over a period of time, changes may take place in the community that lead to the development of a stable or **climax** community. The progression from one stage to the next is known as a **succession** (see Chapter 4).

As an example of succession, a typical area of mown grass, such as a playing field, lawn or park, can eventually develop into woodland, though it may take up to 30 years for the woodland to become established (Figure 5.1). Changes in vegetation occur if there is no mowing or grazing, which would keep the area as grassland. At each of the stages of this succession, there would be populations of invertebrates associated with the vegetation. As cover becomes more dense, small mammals and an increasing number of birds would become evident.

EXTENSION MATERIAL

Succession in an aquatic habitat

In a similar way to the succession described for grassland, from the edge of a freshwater pond or lake into the water, there is a range of plants, each associated with the different depths of water. Animal life amongst the plants depends on the plants and also requires water or wet marshy conditions. If the open stretch of water is left alone for a period of time, plant material, mud and other debris accumulate. Slowly this material builds up from the bottom of the pond or lake, and eventually may rise above the original level of the water. As the habitat becomes less watery, the nature of the plant communities changes, as it goes through the stages of a succession from open water to dry land and eventually to woodland (Figure 5.2). Those plants and animals associated with open water would gradually disappear. Both in the grassland and in the water habitat, if we wish to maintain the range of habitats representing different stages of the succession, some active intervention is required.

Figure 5.2 Profile from open water to edge of pond or lake, representing stages of succession: (1) dry land – can support climax community; (2) bank zone – sometimes flooded; (3) open water

In the open water, some plants are totally submerged, others have leaves on the surface; some float freely, others are rooted on the bottom. In this zone, plants include duckweed (free-floating), water milfoil (submerged), water crowfoot and water lilies. In shallower water, towards the bank, plants include reeds, sedges and grasses, horsetails and perhaps marsh marigold or bogbean. Woody species associated with the banks of wetlands include willows and alder

A key feature of conservation strategies, on a global scale as well as at a local level, is to keep different stages of the succession and prevent the full development of the climax community in the whole of the area. If we wish to retain the range of habitats representing different stages of the succession and the communities associated with these stages, some active intervention and management is required. As you consider the different examples described in this chapter, try to see how the management activities that are adopted interfere with the progress of the succession that would naturally occur.

Conservation in two nature reserves: Bradfield Woods and Wicken Fen

Bradfield Woods

Bradfield Woods is a National Nature Reserve, situated in East Anglia. It occupies an area of about 65 hectares and is surrounded by arable farmland. Woodland has probably existed continuously on the site from the end of the last glaciation some 12 000 years ago. Over the past few hundred years the wood has dwindled to its present size from a much more extensive area. Historical records from the Abbey of Bury St Edmunds indicate that the area has been managed by humans as woodland for at least 750 years.

In the early 1970s, the remaining woodland was on the verge of being destroyed and converted to arable land, but was saved by a small group of local people who saw its potential as a conservation area. Instead of chainsaws and bulldozers moving in, the area has gradually and sympathetically been restored to a working coppice woodland. The scientific value of Bradfield Woods is seen in the richness of its plant and animal life. Its aesthetic value is enjoyed by many visitors, and it is used as an educational resource for adults and children. Volunteers contribute to the conservation programme, and other visitors may be interested in the bird or insect life or just enjoy the spring flowers. In 1994 Bradfield Woods was declared a National Nature Reserve, giving recognition to its value as a conservation site.

A visitor entering Bradfield Woods walks down wide grassy tracks, known as rides, usually with a shallow ditch either side. The wooded areas are laid out in blocks (or fells) and the coppicing or cutting of the woody species takes place in these blocks. **Coppicing** is a method of woodland management that has been practised for many hundreds of years. At intervals, certain tree species (here mostly alder, ash, birch, hazel and sallow) are cut down close to ground level (see *Exchange and Transport, Energy and Ecosystems,* Figures 8.8 and 9.3). New side-shoots then grow out from the stumps (also known as **stools**) and are allowed to grow for, say, 10 to 15 years, then are harvested as poles (Figure 5.3). Different blocks are coppiced in successive years, giving a crop of wood every year. Among the coppiced stools, a few trees (usually oak and silver birch) are allowed to grow to their full size as mature trees. These are known as **standards**. When felled, these trees supply timber, used in building houses. Trees regenerate naturally from seedlings. Some of the ash stools reach a diameter of up to 6 m, with an estimated age of at least one thousand years. They have probably yielded crops of wood continuously over this period. The series of coppiced areas representing the cutting intervals within the 10 to 15-year cycle show successive stages of regrowth of the shoots from the stools. Comparison of the ground flora in these areas reflects stages in a succession, from open ground through to dense shading by the tree cover. This is described in *Exchange and Transport, Energy and Ecosystems,* Figure 9.4. The different stages of the coppicing cycle create a range of habitats and hence a diversity of species.

At least 42 native trees and shrubs have been described from Bradfield Woods, about two-thirds of the total in the whole of Britain. Over 350 species of flowering plants are known from the woods, including a number of rarities. However, if left to grow without any management, the diversity of species within

Figure 5.3 Coppiced woodland in Bradfield Woods, Suffolk. This coppice area has reached a relatively mature stage and is likely to be cut down again within the next 2 or 3 years. Other stages of coppice in Bradfield woods are shown in Exchange and Transport, Energy and Ecosystems, *Chapters 8 and 9.*

BACKGROUND

In the history of Bradfield Woods, the harvested wood has been used for the manufacture of various products, including wooden rakes, handles for scythes, thatching pegs, hazel for daub and wattle (used in local timber-frame buildings), fencing materials and hurdles as well as firewood. Even today, certain wood products are sold, either from the wood itself or (until the 1990s) through a nearby factory.

Species diversity in Bradfield Woods

The open rides offer sites for other flowers and along the edges are shrubs and taller vegetation, providing attractive habitats for insects including butterflies. The rides are relatively dry because of the drainage ditches along the sides. These ditches introduce yet another habitat for a different range of species. As well as the plant life, there are plenty of birds and the abundant birdsong includes that of willow warblers, blackcaps and nightingales. Other animal life includes adders, grass snakes, frogs and toads and a range of typical small woodland mammals. All these are rare or absent in the surrounding farmland.

the wood would gradually diminish. Hazel is an example of a woody species that does not compete well in dense, mature woodland. The rides need to be mown to keep them open and stop invasion of woody species. The shrubby growth at the edges of the rides must be cut back, preferably in stages to ensure continuity of habitats suitable for the insect life.

The descriptions of Bradfield Woods and Wicken Fen (see Additional Material below) illustrate some of the reasons for conservation and the principles behind conservation strategies. These are highlighted as follows:

- Change of land use may lead to loss of diversity in terms of habitats and species (contrast Bradfield Woods with surrounding arable farmland).
- Historically, human activities have already influenced these two sites (people have interacted with the wood or fen, exploiting their productivity).
- Management for conservation requires active interference to maintain diversity by creating a range of habitats (coppicing, mowing, maintaining ditches).
- Scientific interest lies in the range of communities, and rare plant and animal species.
- Nature reserves provide opportunity for academic research and educational study.
- Leisure benefits (medieval rides, present-day visitors) are evident.
- Conservation aims to protect existing ecosystems and maintain genetic resources.

ADDITIONAL MATERIAL

Wicken Fen – a nature reserve in eastern Britain

The original East Anglian fen extended over 3000 km² and developed because drainage was poor and the land was liable to flood. In waterlogged conditions, dead vegetation accumulates and becomes peat instead of decaying. As the peat builds up, the natural succession is to woodland (known as fen carr). For centuries, people exploited the fenland and harvested its products. Around Wicken Fen, they dug peat for fuel, cut reeds and sedges for thatching and gathered vegetation for animal fodder and bedding. From the flooded areas, they caught fish and wildfowl for food. These activities created a range of habitats, from open water through wet reed beds to drier meadows. These habitats supported a diversity of wetland species, both plant and animal. The cutting helped prevent progression to woodland.

Extensive drainage systems were built in the 17th century so that the fenland could be converted into agricultural land. Wicken Fen was one of the few areas that was not drained, and continued to be used in the traditional manner. By the late 19th century, demand for peat, sedge and reeds had largely disappeared and in the early 20th century, very little cutting was done. Gradually shrubs and trees invaded the area.

Today, Wicken Fen occupies about 300 hectares and is owned by the National Trust and managed as a nature reserve. The management strategy aims to encourage and maintain the diversity of species associated with wet fenland communities. Management practices include the digging of peat in a few places to create fresh hollows or pools of water, and ditches and the main water channels (lodes) are kept clear. There is small-scale harvesting of crops by cutting sedge. Meadows and paths are mown to retain this stage of the succession. Different cutting regimes within the year, or on a 2- or 3-year rotation, encourage greater diversity of the vegetation. While a high proportion of the area has developed into woodland (carr), parts are cleared to prevent the woodland stage taking over.

Figure 5.4 Management practices at Wicken Fen help maintain diversity – different stages of succession from open water (ditch in photograph on right) through to woodland. Intermediate stages are maintained by mowing and cutting, and the waterways must be cleared from time to time

The main threat to the fenland is the lowering of the water table. The surrounding farmland now lies several metres below the level of the fen, because the land shrank as it was drained. Water must be pumped up into the fen to maintain the wet conditions, but water gradually seeps out into the farmland below, so parts of the perimeter of the fen have been sealed to try to keep the water in.

Managing grassland and maintaining diversity

Mowing

Mowing is an unselective way of controlling plant growth. When mowing is done by machine, the sward (an area of short grass) is cut to an even height and any unevenness, such as molehills or tussocks, tends to be levelled off. Traditional mowing with a scythe is more sympathetic and retains some unevenness, which can provide useful habitats for invertebrates or ground-nesting birds. The cut vegetation may be removed for use as a hay or silage crop. Removal of the crop means loss of nutrients from the area. In traditional farming systems, nutrients are replaced by dung from animals allowed to graze after mowing, but in modern farming practice, artificial fertilisers are often applied. High levels of nutrients encourage growth of vigorous grasses and these may exclude other less competitive species. If mowings are left on the ground, they may smother certain plant species or prevent seeds reaching the surface and germinating. Accumulated mowings tend to discourage ground-dwelling invertebrates and also create pockets of high nutrient. Mowing in this way can lead to uniformity of species.

Grass grown for silage normally consists of high-yielding varieties and chemical fertiliser is added to maximise the crop. Mowing for silage crops generally starts in late April or early May, followed by two more cuts at 6-week intervals. The emphasis is on high yields, uniformity of growth and mowing at the stage which produces good-quality silage. More diversity is achieved when grass is cut for hay, usually in June or early July. The later cutting date allows more species to flower and produce seeds.

Increased diversity of species can be achieved by varying the season and frequency of mowing and the height of the cutter. Some plants can survive frequent cutting or trampling but many need time to flower and set seed if they are to survive in the area. Table 5.1 gives the flowering period of some grassland species typically found in churchyards. For spring-flowering species, the mowing would be done towards the end of June, but if there is a predominance of species which flower in June and July, the cut should be delayed until late July or August (Figure 5.5).

Generally the even, close cut of frequently mown grassland is inhospitable to insects and other invertebrates, whereas they are encouraged by rough tussocky grass, which has pockets of bare ground or patches of uncut vegetation amongst mown grass. Sometimes an area can be left unmown for 2 to 3 years or even longer. Patches of long grass are important as cover for birds. Reduction in numbers of some birds (such as the corncrake) is probably associated with unsympathetic cutting of grassland. It also helps if fields are cut from the centre towards the outside, so that any birds, particularly those with young chicks, have a chance to keep under cover and escape from the cutter.

Grazing

Grazing animals are selective in their choice of species. In Britain the main domesticated grazing animals are sheep, cattle and horses, though wild animals, particularly rabbits, have considerable influence on the nature of the grassland

BACKGROUND

Agricultural land frequently includes grassland. A successful farmer must maintain productivity at an economic level and this may be seen as the crop removed from the area or through the animals grazing on the land. Elsewhere, grassland is found in parks, golf courses, racecourses, playing fields and gardens, as well as in areas designated for conservation. Maintenance of grassland is achieved through mowing, grazing and sometimes burning, but the profile of plant and animal species present and the diversity achieved depend on the management strategy adopted.

Figure 5.5 Conservation in a churchyard – a Suffolk churchyard in May, where mowing is delayed until after the flowering period as a means of encouraging diversity

ADDITIONAL MATERIAL

Table 5.1 *The flowering periods of some grassland species typically found in churchyards*

Species	Flowering period
lesser celandine	Mar to June
primrose	Mar to May
cowslip	Apr to May
meadow saxifrage	Apr to Jun
greater stitchwort	Apr to Jun
germander speedwell	Mar to Jul
bulbous buttercup	Mar to Jul
bugle	Apr to Jul
meadow vetchling	May to Aug
bush vetch	May to Aug
sorrel	May to Jul
hoary plantain	May to Aug
ox-eye daisy	Jun to Aug
hedge bedstraw	Jun to Sep
lesser stitchwort	May to Aug
lady's bedstraw	Jul to Aug
common knapweed	Jun to Sep
dark mullein	Jun to Oct
burnet saxifrage	Jul to Aug
common toadflax	Jul to Oct

and in preventing the succession to scrub. Sheep bite vegetation close to the ground and thereby maintain a short, even sward. This habitat is favoured by birds such as stone curlews, woodlarks and wheatears. During the day sheep drop their dung across the area, but at night dung is deposited in specific areas, which then become enriched with nutrients which is deleterious to the flora. Trampling by sheep has relatively little effect, except where the soil is loose or on steep hillsides. Cattle wrap their large tongue around the vegetation and, compared with sheep, can consume relatively tall and coarse plants. The resulting sward is more uneven than with sheep because cattle tend to select certain patches. They also trample the ground, leaving hoof marks, bare soil and muddy areas if wet. The term 'poaching' is used to describe the muddy mess made by cattle in this way. Their dung, in the form of cowpats, results in local enrichment of nutrients. Around the cowpat the area is unpalatable, but the cowpat itself provides the opportunity for another community to become established, at least for a short time. Trampling of vegetation and poaching of the ground surface by cattle along the edge of ditches can increase marshy and muddy areas, but if the trampling pressure is too heavy it is harmful. Horses are much more selective than sheep or cattle. They may almost eliminate some species from an area yet ignore others, producing rather patchy vegetation.

In terms of management in a UK farming context, cattle are allowed to graze when the grass is growing and so control the growth of grass. This is from about May through to October. The cattle are then usually housed indoors during the winter. If they were left on the fields over the winter, they would cause damage to the vegetation. Sheep, on the other hand, are kept on the fields throughout the year, although supplementary feeds may also be given from October through to March. Sheep are useful in that, as described above, they maintain a rather short sward. The action of the hooves of both these animals contributes to maintaining small-scale habitats within the grassland. Any stock (including cattle and sheep) must be moved regularly from one field to another to allow for fresh growth of the vegetation and to reduce the build-up of parasites in the field. Goats are also proving to be useful as a management tool for conservation grazing on rough ground, as they nibble woody growth, including shrubs and trees, and thus prevent progression of later stages of the succession. Horses, on account of their rather fussy grazing habits, are less useful in a management strategy aimed at encouraging biodiversity.

The integration of mowing and grazing as conservation strategies is well illustrated by the description of Meadowsweet Fields in Chaseleyfields Farm, given below.

Meadowsweet Fields – an example of management for conservation

Management of hill habitats within a farming structure in a way that enhances biodiversity can be illustrated by a small group of fields on the Welsh borders (UK). This is land that has probably never been ploughed and is still being farmed in the traditional way of 50 years ago. This system entails grazing with low stocking levels of cattle and sheep for part of the year, and taking a hay crop which is used for fodder. In 1998, after some years of neglect, Chaseleyfields Farm was up for sale. There were fears that the land would be swallowed into adjacent farms where intensive systems of frequent ploughing, reseeding, application of fertiliser and herbicides are the norm, thus losing the interesting flora and fauna (including both diversity and a number of rare species). One group of fields, known as Meadowsweet Fields, approximately 11 hectares, was purchased with the determined aim of restoring the traditional system of management and retaining the biodiversity. A little financial help has been obtained from agricultural grants aimed at encouraging more sensitive management of the environment and advice has been obtained from interested conservation bodies. Five years later, the project has progressed and, for the new owner, the rewards have well justified the expenditure and hard physical work.

Mature trees along the field boundaries include oak, ash, wych elm, willows and aspen. The hedges are a mixed array of blackthorn, hawthorn, field maple, hazel, damson, wild apple, elder and holly. A scrub of brambles, dog rose and blackthorn had encroached into substantial areas and was cut back to the original boundaries. Huge overgrown willows have been pollarded before they die. Clearance was never total at a particular time, to ensure that some habitats remained for birds and insects. Overgrown scrub around the edges of the fields was cut back and laid to reinstate the original hedges. These hedges are left wide and only cut on a 3-year rotation to generate good 'wildlife corridors' providing shelter and food for small mammals and birds alike.

The wetter pastures (unsuitable for hay) are grazed during the summer months by beef cattle. Their hooves 'poach' the surface of the ground, maintaining an open structure to the soil and providing holes for seed germination. Cattle also break up the tussocks of rapidly growing tough grasses and prevent them from becoming dominant. A hay crop is taken from the drier fields. To qualify for conservation support grants, there can be no cutting until after 15 July. This gives time for the seeds of annual plants (including grasses) to develop and mature after flowering. The hay is cut with a mechanical scythe, and turned more times than is now usual, to ensure the seeds fall back on to the soil. In the autumn, when reasonable regrowth has formed, there is 'aftermath' grazing. Sheep, being lighter than cattle and nibbling rather than pulling at vegetation as they feed, maintain a better sward for a hay crop, but must be moved off to higher pasture if the ground becomes too wet and in time for the vegetation to grow into the next hay crop.

Each spring and summer month brings a profusion of flowers, peaking in June. The sequence starts in March and by April there are adder's-tongues (ferns) and cowslips, marsh-marigolds in the stream beds, dog's mercury, primroses and bluebells at the woody edges. By May pignut (a particular indicator of old meadow), cuckooflower, wavy bitter-cress, early vetches and ragged-robin are in flower and in June yellow-rattle, heath spotted-orchids, St John's-wort, betony, black knapweed, agrimony, various sorrels, ox-eye daisy, red clover, mints and meadowsweet form colourful swathes across the fields. Annual grasses include crested dog's-tail grass (another indicator of old meadows), timothy, bromes, fescues and quaking grass. Sedges and rushes dominate over grasses in the wetter areas. Thistles, nettles and docks, which are so often conspicuous on badly managed over-stocked pasture, are present but are controlled naturally by competition from the other plants.

Figure 5.6 Meadowsweet fields
(top) Overgrown willows on a field boundary – this shows the initial neglect that had to be gradually and sympathetically restored. If the branches stay on the ground, valuable pasture is lost, the ditches get blocked and the ground becomes waterlogged

(middle) Cows in the meadows play a critical role in the management scheme, through their grazing and also because of the way their hooves 'poach' the ground

(bottom) A view of the meadows, showing its richness of flora during the summer flowering period. Cutting cannot be done until after 15 July in any year, to ensure the flowering period is complete and seeds fall to the ground

So far, more than 200 species of flowering plant have been recorded in these fields, and there are certainly more. This diversity of flora in turn supports a rich fauna, notably insects including many dragonfly species. Ditches and stream banks are maintained, creating a suitable habitat for aquatic and damp-loving species but the modern system of inserting land drains will not be undertaken. Antibiotics and worming agents are not used regularly on the animals, as these would be passed out in dung and spread through the ground, killing natural microorganisms in the soil. To avoid the build up of animal parasites the stock are regularly moved from one field to another. The animals are fed only hay from the meadows and are not given supplementary feeds from outside as this might introduce seeds of undesirable agricultural hybrid grasses. By law, bracken must be controlled; this is crushed or pulled by hand to avoid using chemicals. The odd plant of poisonous ragwort is controlled naturally as it is eaten before flowering, by the caterpillars of the bright red cinnabar moth, which have a safe haven on the land.

In the past five years, neighbouring fields from the original farm have, as predicted, been turned over to more intensive management systems. They are grazed at much higher stocking rates, treated with fertiliser, reseeded and drained. Vigorous grasses out-compete the more rare species; there are no orchids or other flowers so conspicuous in the fields described above. At best, the 'intensive' fields probably support around 20 species of flowering plants, including woody hedge plants. The profits may be higher, but the biodiversity has been dramatically impoverished.

Scrub clearance

Scrub is a stage of the succession between grassland and woodland. In the UK, typical scrub in lowland areas includes hawthorn, buckthorn, sometimes blackthorn (particularly in ancient woodland), wild rose, elder, wayfaring tree, willow, sallow, maple, birch and gorse. These are woody, often thorny, species. The actual range of species is influenced by the nature of the soil and other environmental factors. Scrub may be invasive in neglected grassland, derelict urban sites, inland from coastal areas such as dunes, close to waterways, heaths and in woodland. Mechanical clearance is often undertaken to help maintain the more open stage of the succession and sometimes burning or herbicides are used. Regular cutting back of scrub creates a habitat similar to coppice. Scrub does, however, have a conservation value in its own right. As an example, hawthorn or other dense scrub can offer suitable habitat for insect life and it is also favoured by some birds, notably warblers and nightingales. To encourage diversity of species, in some areas management may be by a mixture of grazing and cutting in patches. This helps to create a mosaic of vegetation which is sympathetic to the wildlife associated with scrub, as well as maintaining the essential elements of the open habitat.

Use of fire

Deliberate burning used as a management tool has both advantages and disadvantages. It provides a way of controlling shrubby growth and other coarse vegetation, usually in locations where it would be difficult to use machinery for cutting and clearing, or which otherwise would be very labour intensive to keep clear. In the UK, it is practised mainly in upland moorland communities, which

> ### QUESTION
>
> Cowpats in the conservation meadows usually disappear within 2 or 3 weeks, whereas those in neighbouring intensively farmed fields may take 6 months or more to disappear.
> - Try to think of some reasons that would explain this difference.

are managed for sheep, deer and red grouse. These are areas dominated by dwarf shrubs, such as heather and bilberry. Sheep find spring growth of purple moor grass and young heather particularly palatable and red grouse feed mainly on heather. The fire helps stimulate the growth of young heather and checks the growth of young shrubby seedlings and coarse herbaceous vegetation. This helps prevent the succession progressing to scrub from open heathland.

Ideally, the burning should be carried out in patches, to create an irregular mosaic pattern with vegetation of different ages. This encourages a range of habitats, which in turn support a diversity of species, including invertebrates. After an area has been burnt, perennial plants regenerate from the underground parts and invertebrates can recolonise from neighbouring areas. Burning should not be done too frequently and certainly not every year – often it is done on a 10- to 15-year cycle, by which time the mature heather and other shrubby plants have reached a height of well over 30 cm. Inevitably there is damage, particularly to fauna, such as reptiles and molluscs, which cannot escape. Under certain conditions, the intense heat of the fire can have adverse effects on the soil, and moss and lichen cover may be destroyed. Burning would be unsuitable in areas colonised by bracken since this encourages new growth and undesirable spread of the bracken. To minimise damage, burning is permitted at certain times of year only – in the UK this is during autumn or early spring, the official season being 1 October to 15 April. During this period, at least some animals are overwintering in sites that are protected from fire, and there is less risk of fires burning out of control in the very dry conditions of a hot summer. Overall effects of burning are complex, but traditionally it has been a useful way of managing certain areas. Provided the strict regulations relating to burning are enforced, and when combined with grazing, burning will continue to be a valuable conservation strategy in a limited range of habitats (Figure 5.7). These include restoration of reed beds and controlling scrub, as well as the moorland heath described above.

Figure 5.7 Controlled burning of heather as a management tool on Exmoor, western England

Coppicing

The practice and role of coppicing in the management of woodland is highlighted in the description of Bradfield Woods (pages 65 to 66). Further information is given in *Exchange and Transport, Energy and Ecosystems*, Chapters 8 and 9. In the descriptions, emphasis is given to the way that biodiversity is both maintained and increased through the range of habitats that can be found within a coppice woodland.

Integrating conservation and food production

This section discusses how intensive food production has affected wildlife and the examples that are described should help you to understand how modern farming practice can be manipulated and adapted to enhance biodiversity. Earlier in this chapter on conservation, we have looked closely at some British habitats and seen how, through careful management, a diversity of species can be maintained. Specially protected areas, like Bradfield Woods, Wicken Fen and many other reserves, are very valuable in conservation terms but they occupy only a tiny fraction of the land (probably less than 2 per cent). There is, however, considerable scope for integration of conservation with current

CONSERVATION

Table 5.2 *Changes in farming practice (from traditional small-scale to intensive large-scale systems) that have resulted in loss of habitat and of species diversity*

Changes in farming practice	Effect on habitat and species
uncropped land has been converted to cropped land	• reduced vegetation diversity • fewer invertebrates, small mammals, birds
large woods fragmented to small woods	• decline of woodland species
hedges removed (increased size of fields)	• decline in small mammals, owls, overwintering predatory invertebrates • removal of berry and nut food source for winter thrushes and other birds
ponds filled in	• less aquatic vegetation • decline in newts, frogs, aquatic invertebrates (e.g. dragonflies)
permanent pasture converted to temporary ley (grassland)	• reduced diversity of vegetation • decline in earthworms and other invertebrates • fewer birds such as skylarks, lapwings, thrushes, gulls, golden plovers
mixed crops replaced by arable monoculture	• reduced invertebrate diversity • decline in birds such as lapwings, skylarks, stone curlews
crop rotation replaced by same crop each year	• reduced diversity of invertebrates including butterflies • reduced diversity of birds
undersowing with grass or clover replaced by no undersowing	• fewer summer invertebrates • nitrogen inputs needed on land
application of herbicides, insecticides, fungicides	• direct toxicity • reduced invertebrate abundance and diversity, effects through food chains on food supplies

agricultural practices in ways that have benefits in terms of encouraging species diversity.

In Britain, demand for increased food production over the past 50 years has meant that traditional small-scale farming systems have largely been replaced by intensive large-scale systems, which rely heavily on machinery, fertilisers and pesticides (including herbicides, fungicides and insecticides). In addition, more land has been brought into cultivation by clearing woodland, draining wetlands and ploughing up meadows. Table 5.2 lists a range of farming practices that have resulted in loss of 'natural' habitat, with an associated reduction in species diversity.

Even on intensively farmed land, there are ways of improving the diversity of wildlife, often with minimal economic loss in terms of crop yield. Areas with semi-natural habitats can be increased and diversity can be created within other habitats. The margins of cropped areas can be valuable provided the area is large enough for species to become established. Any such areas must also be managed in a way that encourages species diversity.

The description of Meadowsweet Fields (pages 68 to 70) gives a detailed account of how conservation can be integrated successfully with production on farmland in a particular situation. Some more general farm management practices and how these can be modified to benefit conservation are summarised in Table 5.3. A few points are considered in more detail below.

QUESTIONS
- List some of the benefits of maintaining species diversity and give some definite examples.
- Think about the importance of maintaining genetic resources as well as the aesthetic and scientific value of management for conservation.

Table 5.3 *Some farming practices and ways that management for conservation can be integrated within the farm system*

Habitat	Farm use and management practices	Management for conservation	Benefits
grassland (pasture)	• used for grazing, hay, silage • managment involves weed control, fertiliser and lime, farmyard manure, reseeding, adjustment of stocking density	• use minimum amounts of fertiliser and herbicide • retain damp areas and corners • avoid reseeding • prevent colonisation of scrub and avoid planting trees where flora is rich	important for plants, butterflies and other insects, birds including winter thrushes, waders
woods, copses	• used for timber, firewood, game conservation • management involves felling and replanting, coppicing and thinning in rotation; maintenance of rides and glades	• plant or retain native species • manage part at a time to allow recolonisation • use coppice management on rotation where suitable • maintain wide rides, cutting alternate sides each year	important for plants, butterflies, birds, mammals
hedges	• used as stock-proof fences, for shelter and game conservation • managed by laying, cutting, coppicing, annual trimming	• maintain hedges to be thick at the base and of reasonable height • keep some trees in the hedge • laying or coppicing at suitable intervals	important for plants, butterflies, birds, mammals
water courses	• useful as barriers for stock and for drainage and irrigation • need regular cleaning, involving mechanical and chemical weed control	• manage part at a time to allow recolonisation • keep gentle profiles • ensure water is present throughout the year	important for plants, dragonflies, fish, birds, mammals
ponds	• used for drinking by stock, for irrigation, angling, shooting	• manage part at a time to allow recolonisation • make sure water remains unpolluted throughout the year • avoid too much shading by surrounding trees	important for plants, dragonflies, fish, amphibia, birds, mammals
lanes, roadside verges	• needed for access to fields and buildings • managed by mowing, herbicides	• cut late, after flowering and remove cuttings if possible • avoid herbicides	important for plants, insects including butterflies

Minimise use of pesticides – see Chapter 4 (pages 57 to 61) for a discussion of strategies than can avoid loss of insect life through use of insecticides. Methods used to increase the prevalence of natural predators in effect increase biodiversity within the farming ecosystem.

Organic farming methods – Many of the strategies adopted by organic farmers aim to achieve an equilibrium within the ecosystem, rather than to dominate the land with a monoculture that utilises chemicals in the form of pesticides and herbicides to increase yields. Organic methods usually encourage diversity in the plant and animal communities found close to and within the crops. To some extent, this biodiversity can be exploited in practices aimed at control of pests within the crops and to increase crop yields. Some cultivation methods adopted by organic farmers are described in Chapter 4 (pages 58 to 59).

Conservation 'headlands' – in the UK, as part of the 'Countryside Stewardship Scheme' there is encouragement to integrate strips of unploughed land into arable farmland and to manage these strips in a way that increases biodiversity. These strips may appear in different forms, say as a 2 m or 6 m strip (known as a 'headland') around the edges of a crop, or as a beetle bank across a cultivated field (see Chapter 4, page 58). The headlands are sown with suitable 'conservation seed mixtures' to include plants that encourage a range of species of insects or other animals, especially invertebrates. There are strict regulations regarding the management of these strips, covering the regime for mowing, which insecticides are permitted and when they may be used, and the times at which herbicides may be applied. The insects, weeds and seeds that are associated with the strips provide food sources for many of the farmland and game birds. The flowers provide nectar and pollen for butterflies, predatory and other pollinating insects and the habitats created may allow rare arable flowers to flourish. (Note that the information given here is correct for the years 2001 to 2003, and while details of schemes may change in future years or be different in other locations, the principles that are being applied are appropriate to illustrate how conservation can be integrated with farming.)

Financial incentives – Farmers, like any other business managers, must make a profit out of their activities. They may also need advice as to how to implement conservation and encourage diversity of species on their land. A number of schemes are in operation which provide the opportunity for creation of wildlife habitats, or direct funding, often with advice to farmers.

Government incentives are given to farmers to encourage them to enhance wildlife on their farms. These incentives are provided by schemes run in the UK by the Department for Environment, Food and Rural Affairs (DEFRA), and previously by the Ministry of Agriculture, Fisheries and Food (MAFF). As an example, during the year 2000, a pilot scheme with the name 'Arable Stewardship' was operating in specified areas in East Anglia. The scheme offered payments to arable farmers if they adopted certain management practices to encourage wildlife on their farms. Several options were available and, in making an application for the grant, a farmer would have to satisfy a number of conditions. One option was to promote overwintering of stubble (from cereal or linseed crops) until July of the following year. Only limited use of herbicide was permitted and there were other management conditions. This aimed to provide foraging and breeding sites for birds (such as grey partridge, corn bunting and the skylark) and give suitable conditions for the brown hare. It is also likely to provide breeding sites for lapwing and stone curlew, and to encourage plant species associated with arable areas that have become rare. Similar benefits would be achieved with a second option, namely the undersowing of cereal crops with a grass / legume (ley) seed mix. In this practice, after harvest of the main crop, the undersown ley must be retained until July of the following year, though the land may be grazed with cattle or sheep, provided the stocking density is not too high. Other options included under this grant scheme include the development of beetle banks (see page 58) and uncropped wildlife strips which allow rare annual and biennial plants associated with arable land to become established and set seed. This seed is valuable as food for seed-eating birds. Grants are available also for maintenance and restoration of hedges – by laying, coppicing and planting with appropriate native species to fill gaps. To quote examples of payments in this scheme, for beetle banks a farmer is likely to receive about £600 per hectare (or £12 per 100 metres) or £525 for maintaining uncropped wildlife strips. This Arable Stewardship Scheme is operated under the 'EC Rural Development Regulation (1257/199)'. As a pilot scheme, the benefits are being carefully monitored and participants are expected to make a commitment lasting for at least 5 years.

Summary

Finally, as an oversimplification, we summarise below the choices in terms of land use, its management and the maintenance of plant and animal species diversity in the interests of conservation:

- progression or succession to natural climax community
  → less species diversity
- intervention in a succession and management for conservation
  → more species diversity
- intensive agriculture → less species diversity
- integration of conservation with farming practice
  → more species diversity.

The European Union Habitats Directive

EU nature conservation policy is based on two main pieces of legislation: the **Birds Directive** and the **Habitats Directive**. The priorities of the Habitats Directive are to create a European ecological network of special areas of conservation, known as **Natura 2000**, and to integrate nature conservation requirements into other EU polices, including agriculture, regional development and transport.

The Council of the European Communities Directive (92/43/EEC, 21 May 1992) on the conservation of natural habitats and of wild fauna and flora, states that the main aim is:

> *'to promote the maintenance of biodiversity, taking account of economic, social, cultural and regional requirement'.*

The Directive recognises that, within the EU, natural habitats are continuing to deteriorate and that an increasing number of wild species of plants and animals are threatened as a consequence. The Directive also defines a range of natural habitats and species of plants and animals which should be given priority in order to conserve them. **Annex I** of the Directive gives a list of the natural habitat types whose conservation requires the designation of **special areas of conservation**. This includes:

- coastal habitats, including estuaries, saltmarshes and shingle beaches
- coastal sand dunes
- freshwater habitats
- temperate heath and scrub
- natural and semi-natural grassland formations

- raised bogs, mires and fens
- rocky habitats, including screes, and caves
- forests, including old sessile oak woods and yew woods of the British Isles.

Annex II of the directive includes a list of animals and plants which are recognised as being endangered, rare or requiring particular attention. Each member State of the EU is required to contribute to the creation of Natura 2000 by identifying natural habitats listed in Annex I and habitats of the species listed in Annex II within their region. It is their responsibility to establish the necessary conservation measures and appropriate management plans to maintain these habitats in order to ensure biodiversity through conservation of the natural habitats and wild fauna and flora.

QUESTION

Bradfield Woods and Wicken Fen are two nature reserves, and their descriptions in this chapter illustrate conservation in practice. Similar principles would be applied to many other habitats where the aim is to maintain a diversity of species. Use these questions to compare these two reserves (or compare one with another nature reserve that may be familiar to you).

- What human activities over the past 500 years or more have influenced the vegetation in the area? What would be the natural climax vegetation in the area? In what ways has the productivity of the area been exploited and harvested by local people? What sort of land surrounds the reserve today? How does the reserve differ from the surrounding land in terms of diversity? What management strategies are being used to help maintain the diversity within the nature reserve?
- Suppose you were involved with a group of people interested in creating a small nature reserve out of a piece of wasteland. Draw up a list of principles that you would wish to apply to make it an attractive conservation area. Then give examples of some of the practical things you would do to create and maintain the reserve.

Genes, alleles and sources of new inherited variation

Variation

When organisms reproduce, their offspring are of the same species and the members of one family are all similar to one another and to their parents in their specific characteristics. We recognise the species characteristics of humans as being different from those of rabbits and gorillas, but we also recognise that the individuals within one species vary in small ways. These small differences constitute **variation**, which may be the result of genetic differences, or of the influence of the environment, or a combination of both. In some cases it is difficult to determine what contribution is made by heredity and what is due to the environment, especially if the differences are very small. In humans, factors such as colour of the skin, hair colour, weight, shape of head and facial features all show variation and we know that many of these are inherited characteristics because we use them to distinguish between the various ethnic subgroups of our species. Some of these factors, for example weight, can be affected by the level of nutrition or exercise, which are both environmental influences.

We recognise two forms of variation:
- discontinuous
- continuous.

In **discontinuous variation**, there are clear-cut differences in the forms of the various characters, determined by the alleles of a single gene. The characters in the garden pea (*Pisum sativum*), which the Austrian monk Gregor Mendel worked with when he did his experiments on inheritance, are good examples of discontinuous variation. When tall pea plants were crossed with dwarf pea plants, all the plants were tall. In subsequent generations, which had been allowed to self-pollinate, both tall and dwarf plants appeared, but there were no intermediates (Figure 6.1a). Similarly, there are polled (hornless) and horned forms of cattle, vestigial- and normal-winged *Drosophila* (Figure 6.2) and yellow and purple seeds in maize. The differences between the forms are determined by alternative forms, or **alleles**, of a single gene and no measurements are needed to distinguish between them. Examples of single gene inheritance in humans include tongue-rolling, lobed and non-lobed ears and ABO blood groups.

Much of the variation that occurs within a species is concerned with size, mass or shape, characters whose variation does not fall easily into groups. Such characters are usually determined by several genes, each of which has a small effect. It is difficult to measure the individual contribution of each gene, especially as the environment has a great influence on this type of variation. This variation is known as **continuous variation** and is clearly shown by height in humans (Figure 6.1b). Within a human population, there is usually a range of height in adult males from 150 to 190 cm. There are no separate categories but a continuous distribution of values. If we were to measure all the adult males in an area, group them into classes that differed from each other by

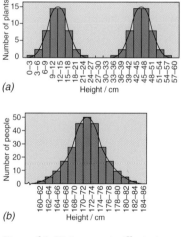

Figure 6.1 Histograms to illustrate: (a) discontinuous variation in height of pea plants, some of a dwarf variety and some of a tall variety, showing that there is no overlap in height distribution; (b) continuous variation in the height of adult male humans, showing a normal distribution about the mean

(a)

(b)

Figure 6.2 Adult fruit flies
(Drosophila melanogaster) *with*
(a) normal-winged (wild-type) and
(b) vestigial-winged phenotypes
(appearances)

2 cm and plot a distribution curve, we would obtain a normal distribution. We would probably find that most values would be around 170 cm, the mean value. An individual can be measured and classified using features that vary continuously.

In the early years of genetics, it was difficult to determine to what extent such characters as height or mass, which show this type of continuous variation, are controlled by genes or are the effects of environmental factors.

BACKGROUND

A Danish geneticist, Johannsen, working in the early 1900s, designed a series of experiments to find out whether seed mass in the dwarf bean (*Phaseolus vulgaris*) was inherited or varied due to environmental factors (Figure 6.3). He collected a number of 'pure lines' of seeds, each line consisting of the descendants of one seed. The seeds in each line were genetically identical and each line had a different mean seed mass. In one experiment, he mixed seeds from all these lines together and then selected a sample of large seeds and a sample of small seeds. He grew these and collected the seeds from the offspring of each sample and determined the mean seed mass. He found that the mean seed mass produced from the offspring of the larger seeds was

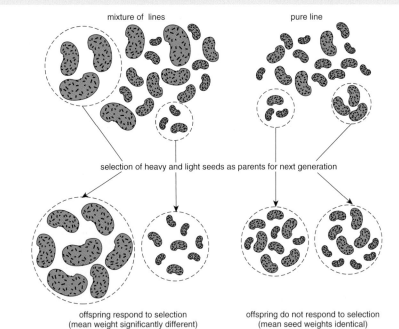

mixture of lines   pure line

selection of heavy and light seeds as parents for next generation

offspring respond to selection   offspring do not respond to selection
(mean weight significantly different)   (mean seed weights identical)

Figure 6.3 Johannsen's experiment with the dwarf bean showed that genes control measurable characteristics and that it is possible to distinguish variation caused by genetic and environmental factors

greater than the mean seed mass from the offspring of the smaller seeds. He compared these results with another experiment in which he selected a sample of large seeds and a sample of small seeds from the same pure line. When these were grown and the seeds from the offspring collected, both samples had the same mean seed mass.

When Johannsen selected from the mixture of lines, he would have picked out seeds that were genetically different, but when he selected from one line, they were genetically identical and the differences in size were due only to different environmental influences. His experiment showed that measurable variation was controlled by genes and that it was possible to distinguish between hereditary factors and environmental ones. Johannsen did not relate the difference to any particular genes.

Later experiments on grain colour in wheat showed that a number of different genes contribute to characters which show continuous variation and the **multiple gene hypothesis** was put forward. In wheat, the grains vary in colour from white to red and this character is now known to be controlled by two genes, which have a cumulative, or additive, effect.

The term **polygenes** is now used instead of multiple genes and **polygenic inheritance** describes this type of inheritance. Its main features are that:
- the characters are controlled by two or more different genes
- each individual gene has a small effect
- there is considerable environmental influence
- the characters show a continuous range of variation.

Monohybrid inheritance

Before we consider how characters are inherited, it is important to become familiar with some of the terms used in the study of inheritance. We already have a definition for a **gene** as a specific length of DNA coding for a polypeptide. It occupies a specific site on a chromosome, called a **locus**.

An **allele** is an alternative form of a gene occupying the same locus on a chromosome as other alternative alleles of the same gene. There are at least two forms of each gene. For example, in Mendel's experiments on peas, he investigated two alleles of the gene for stem length: tall and dwarf. When homologous chromosomes pair up during prophase I of meiosis, these alleles would occupy corresponding positions opposite each other (Figure 6.4).

The alleles are responsible for determining contrasting traits of a character; in this case the character is stem length and the traits are tall (represented by T) and dwarf (represented by t).

The term **genotype** is used to describe the genetic constitution of an organism for the character under consideration. It is represented by the alleles present, so a tall plant would have the genotype **TT** or **Tt**, and a dwarf plant **tt**.

The **phenotype** is the appearance of the character in an organism. It results from an interaction between the expression of the genotype and the effect of the environment. Mendel's pea plants appeared tall or dwarf.

In a diploid organism the alleles occur in pairs, one on each homologous chromosome. There is usually a **dominant** allele, which influences the appearance of the phenotype even if it is present with the alternative allele. It is the convention to represent this dominant allele with a capital letter. The **recessive** allele can only influence the appearance of the phenotype if the dominant allele is not present. The recessive allele is represented by a lower case (small) letter. In the case of the peas, the allele for tall is dominant to the allele for dwarf, so plants with the genotype **TT** or **Tt** will appear tall but **tt** will be dwarf.

If both alleles at a given locus are the same, the organism is said to be **homozygous** for that particular character. **TT** indicates that a pea plant is homozygous for tallness, and **tt** that it is homozygous for dwarfness. If the

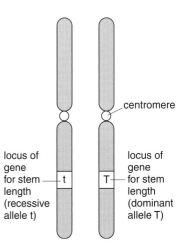

Figure 6.4 A pair of homologous chromosomes, showing that different versions (alleles) of a gene occupy the same position (locus) on a chromosome

79

GENES, ALLELES AND SOURCES OF NEW INHERITED VARIATION

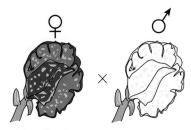

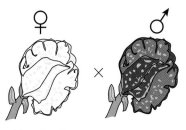

(a) pollen from
white-flowered plants used to
pollinate purple-flowered plants

(b) pollen from
purple-flowered plants used to
pollinate white-flowered plants

= female   = male

*Figure 6.5 Mendel carried out
reciprocal crosses in his breeding
experiments with peas in order to
show that the sex of the plant did not
affect the outcome of the experiment*

alleles at a given locus are different, then the organism is said to be **heterozygous** for that particular character. A plant with the genotype **Tt** is be heterozygous for tallness.

Monohybrid cross

A monohybrid cross involves a single character that is controlled by one gene with two or more alleles. Mendel, whose experiments on the inheritance of characters in the garden pea (*Pisum sativum*) led to the formulation of two fundamental laws of genetics, carried out a large number of monohybrid crosses. He had identified a number of distinct varieties of the garden pea, each showing contrasting traits of a particular character, such as tall or dwarf stem length, axially or terminally positioned flowers and purple or white flower colour. He selected pure-breeding plants with the contrasting traits and cross-fertilised them, producing **hybrids**.

If we take flower colour as an example, Mendel had to be sure that the parental types, one with purple flowers and one with white flowers, were pure-breeding. That is to say, when allowed to self-pollinate, they only produced offspring with one type of flower colour, generation after generation. He then removed the stamens from plants with purple flowers and dusted pollen from white flowers onto their stigmas. He also carried out the reciprocal cross, in which pollen from plants with purple flowers was dusted onto the stigmas of white flowers from which the stamens had been removed (Figure 6.5).

The seeds of all the plants were collected and planted the following year. When these plants, known as the **first filial generation** (or F_1), flowered, all the flowers were purple. This suggested to Mendel that purple flower colour was dominant to white. He allowed the flowers in the F_1 generation to self-pollinate, collected the seeds and planted them the following year. In this, the **second filial generation** (or F_2), there were both purple-flowered plants and white-flowered plants, but the ones with purple flowers were three times as frequent as those with white flowers, a ratio of 3 : 1. Mendel went on to analyse the offspring of this F_2 generation by allowing these flowers to self-pollinate (Figure 6.6). He found that the white-flowered plants of the F_2 only gave rise to white-flowered plants in the **third filial generation** (or F_3), but one third of the purple-flowered plants produced offspring with purple flowers and two-thirds produced a mixture of plants with purple flowers and plants with white flowers. The ratio of purple-flowered plants to white-flowered plants was, again, 3 : 1.

Mendel investigated a number of different characters with one pair of contrasting traits and always found very similar results and ratios. From such experiments, he worked out his **principle of segregation**, in which he suggested that the characteristics of an organism are determined by factors that occur in pairs, and only one of a pair of factors can be represented in a single gamete. He thought of the units of inheritance as 'particles', which were passed from generation to generation. He recognised that sometimes their effects were hidden, but that the particles themselves were passed on unchanged.

We can explain what is happening in the monohybrid cross in terms of modern genetics (Figure 6.7). Because the parent plants are homozygous for flower

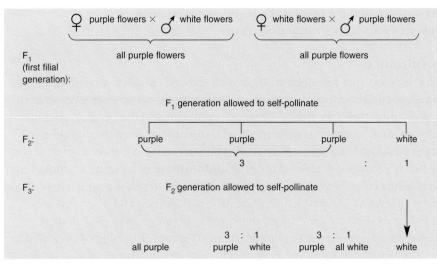

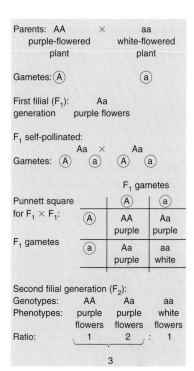

Figure 6.6 In a monohybrid cross with pure-breeding parent plants, when two of their offspring (F₁ generation) are allowed to self-fertilise, the ratio of phenotypes in the F₂ generation is 3 : 1. Further self-pollination of the F₂ plants to produce the F₃ generation will indicate which of the F₂ plants were pure-breeding.

colour, they each produce only one type of gamete. Fertilisation results in a heterozygous hybrid F_1 containing one dominant allele and one recessive allele, so the phenotype is purple. As the F_1 plants are all heterozygous, when gametes are formed, half of them will carry the dominant allele and half the recessive. Fertilisation is random, so the possibilities are that about $\frac{1}{4}$ of the offspring will be homozygous dominant (**AA**), $\frac{1}{4}$ will be homozygous recessive (**aa**) and $\frac{1}{2}$ will be heterozygous (**Aa**), giving a genotypic ratio of 1 : 2 : 1. As **A** is dominant to **a**, then the phenotypic ratio will be 3 purple : 1 white.

The events of meiosis explain how the alleles are separated. The inheritance of certain characters in animals follows the same pattern of inheritance as Mendel discovered in the pea plants (see *Practical: A breeding experiment using Tribolium*). In guinea pigs, the allele for short hair (**H**) is dominant to the allele for long hair (**h**). A pure-breeding short-haired guinea pig (**HH**) was mated with a pure-breeding long-haired guinea pig (**hh**). All the offspring in the F_1 had short hair. A male and a female from this generation were allowed to interbreed (Figure 6.8). They produced offspring (the F_2), some with short hair and some with long hair, in the ratio of 3 short hair : 1 long hair.

In this example, the offspring that have short hair are either homozygous, (**HH**) or heterozygous (**Hh**). The genotype of the long-haired offspring must be **hh**, the homozygous recessive. In order to discover the genotype of one of the short-haired offspring, it is necessary to carry out a **test cross**, or **back cross** (Figure 6.9). This involves mating the guinea pig of unknown genotype with a long-haired guinea pig, which must be **hh**.

If the unknown genotype is **HH**, then the offspring of the test cross will always have short hair, because they will inherit a dominant allele from that parent. If the unknown genotype is **Hh**, the offspring can inherit either **H** or **h**. If any of the offspring have long hair, then we know that the unknown genotype must be **Hh**. The ratio of the offspring in such a test cross, where one parent is

Figure 6.7 Convention for describing a genetic cross, showing that the possible combinations between gametes can be expressed in a Punnett square. The events of meiosis explain how the alleles (A and a) are separated to form the gametes.

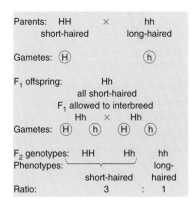

Figure 6.8 Results of a monohybrid cross between a homozygous (HH) short-haired guinea pig and a homozygous (hh) long-haired guinea pig, followed by interbreeding of the F₁ offspring

If unknown genotype is homozygous (HH):

HH × hh (homozygous recessive)

Gametes: (H)   (h)

Offspring: Hh

all short-haired

If unknown genotype is heterozygous (Hh):

Hh × hh

Gametes: (H) (h)   (h)

Punnett square

	(H)	(h)
(h)	Hh	hh

Offspring: short-haired : long-haired

1   :   1

*Figure 6.9 Results of the test crosses with long-haired (**bb**) guinea pigs, used to distinguish between short-haired homozygous (**HH**) and short-haired heterozygous (**Hb**) individuals*

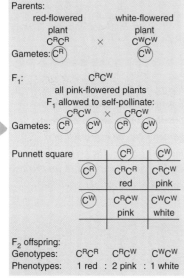

Where neither allele is dominant, they are both shown using upper case letters, in this case R for red flower colour and W for white flower colour

Parents:

red-flowered   white-flowered

plant   plant

C^RC^R × C^WC^W

Gametes: (C^R) (C^W)

F_1: C^RC^W

all pink-flowered plants

F_1 allowed to self-pollinate:

C^RC^W × C^RC^W

Gametes: (C^R) (C^W) (C^R) (C^W)

Punnett square

	(C^R)	(C^W)
(C^R)	C^RC^R red	C^RC^W pink
(C^W)	C^RC^W pink	C^WC^W white

F_2 offspring:

Genotypes: C^RC^R C^RC^W C^WC^W

Phenotypes: 1 red   :   2 pink   :   1 white

*Figure 6.10 Codominance in snapdragons (*Antirrhinum *sp.), showing how it leads to the appearance of a new phenotype in the heterozygous condition*

heterozygous and the other is the homozygous recessive, is always 1 heterozygous : 1 homozygous recessive.

Dominance

The characters that Mendel investigated all showed complete dominance, where one allele influences the appearance of the phenotype, even if it is present with an alternative allele. So, the homozygous dominant and the heterozygote both have the same phenotype and cannot be distinguished from one another. In some cases, the heterozygotes appear to be intermediate between the two homozygotes. This is known as **codominance**. It can be illustrated by the inheritance of flower colour in snapdragons (*Antirrhinum* sp.), where there are two alleles, red and white. Neither allele is dominant and when red and white-flowered plants are crossed the heterozygote produces pink flowers, intermediate between the two homozygotes (Figure 6.10). Self-pollination of the pink heterozygotes of the F_1 results in flowers of all three colours, in the ratio 1 red : 2 pink : 1 white, in the F_2.

The genotypic ratio of 1 : 2 : 1 is typical of a monohybrid cross, and in this case the phenotypic ratio is the same because the heterozygote colour is not the same as the two homozygotes.

There are codominant alleles in both the ABO and MN blood group systems, which will be discussed later in this chapter. Codominance is also shown in the inheritance of the mutant allele that codes for the production of an abnormal β-globin chain in haemoglobin. This allele (**Hb^S**) codes for the amino acid valine instead of glutamic acid on the β-globin chain and results in the formation of haemoglobin S, responsible for the condition known as sickle-cell anaemia. If an individual is homozygous for this allele (**Hb^S/ Hb^S**), the solubility of the haemoglobin is low and it crystallises, causing the red blood cells to be distorted into a crescent or sickle shape (Figure 6.11). Many red blood cells are destroyed and the individual suffers from sickle-cell anaemia, a condition that may prove fatal. In heterozygous individuals, **Hb^S/ Hb^A**, where **Hb^A** is the allele for normal haemoglobin, half the haemoglobin is affected and half is normal. In these individuals, the red blood cells are not usually affected, but if their blood is kept in conditions of low oxygen concentration and the red cells are observed, some will be seen to be sickle-shaped. In this case, both alleles contribute to the phenotype and the heterozygotes can be identified by looking at the red blood cells.

Multiple alleles

When there are three or more alleles for a specific locus, a gene is said to have multiple alleles. In any diploid organism, a gene can only be represented twice, that is, there is one allele on each corresponding locus of the homologous chromosome pair concerned. Where there are multiple alleles, the number of phenotypes will depend on the number of alleles and what sort of dominance is shown.

In humans, the ABO blood group system involves three different alleles. The gene locus is represented by I (for isohaemagglutinogen) and determines which form of antigen is present on the surface of the red blood cells. The three alleles are **I^A**, **I^B** and **I^O**. **I^A** and **I^B** are codominant and code for slightly different antigens, but **I°** is recessive to both **I^A** and **I^B** and produces no antigens. An individual carries two alleles, so there are six possible genotypes and four possible phenotypes (Table 6.1).

From the table, it can be seen that individuals with blood group A can either be homozygous for **I^A** (**I^A I^A**) or heterozygous (**I^A I°**). Similarly, individuals with blood group B can either be **I^B I^B** or **I^B I°**. Individuals with blood group AB are heterozygotes (**I^A I^B**), whilst those with group O must be homozygous recessive (**I° I°**). An understanding of this system enables the blood groups within families to be worked out and is of practical importance in relation to blood transfusions (Figure 6.12).

Another example of a multiple allele system is seen in the gene for the determination of coat colour in rabbits, where there are four alleles:
- agouti (**C**) – the colour of the fur of the wild rabbit; each hair has a grey base, a yellow band and a black tip
- chinchilla (**c^ch**) – silvery, grey-coloured fur; the hairs lack the yellow band
- himalayan (**c^h**) – a white coat except for black nose, ears, feet and tail
- albino (**c**) – no pigment; the fur is pure white.

Development of the pigment in the fur of the extremities of the himalayan rabbit is due to a temperature-sensitive enzyme that only exerts its effect below a certain temperature. Baby himalayans are pure white and only develop their dark colouring when they leave the nest. This is an example of an interaction between the genotype and the environment resulting in the production of a phenotype.

The dominance sequence is **C** > **c^ch** > **c^h** > **c**. There are ten possible genotypes but only four possible phenotypes (Table 6.2).

Dihybrid inheritance

Following his experiments on monohybrid inheritance, Mendel carried out a second set of experiments to investigate the inheritance of two pairs of contrasting characters. He made dihybrid crosses between pure-breeding pea plants differing in two characters. From his monohybrid crosses, he knew that the trait for round seeds (**R**) was dominant to that for wrinkled seeds (**r**), and that the trait for yellow cotyledons (**Y**) was dominant to that for green cotyledons (**y**). So he crossed plants with round seeds and yellow cotyledons (**RRYY**) with plants that had wrinkled seeds and green cotyledons (**rryy**). The F_1 generation all showed the dominant characters of round seeds and yellow cotyledons. When these were self-pollinated, the F_2 generation had four different phenotypes: two that resembled the parental types and two new types, combining characters of both parents (Figure 6.13). The phenotypes and the ratios of plants obtained in the F_2 were as follows:
- 9 (round seeds and yellow cotyledons)
- 3 (round seeds and green cotyledons)
- 3 (wrinkled seeds and yellow cotyledons)
- 1 (wrinkled seeds and green cotyledons).

Mendel observed that the ratio of round seeds to wrinkled seeds was 3 : 1 and that the ratio for yellow to green cotyledons was also 3 : 1, the same ratios as

Figure 6.11 Scanning electronmicrograph of red blood cells from an individual suffering from sickle-cell anaemia, showing that some of the cells are misshapen as a result of a mutant allele for one of the polypeptide chains in haemoglobin

Table 6.1 *ABO blood group genotypes and phenotypes*

Genotype	Phenotype
I^A I^A	A
I^A I°	A
I^B I^B	B
I^B I°	B
I^A I^B	AB
I° I°	O

Figure 6.12 Genetic cross between a blood group A individual and a blood group B individual, illustrating that all four blood group phenotypes are equally possible in their offspring

GENES, ALLELES AND SOURCES OF NEW INHERITED VARIATION

Table 6.2 *Table of phenotypes and genotypes of rabbits*

Phenotype	Genotype
agouti	CC, Cc^{ch}, Cc^h, Cc
chinchilla	c^{ch}c^{ch}, c^{ch}c^h, c^{ch}c
himalayan	c^{h}c^h, c^{h}c
albino	cc

appeared in the monohybrid crosses. He saw that the two 3 : 1 ratios were associated in the same cross and deduced that the two pairs of characters behave quite independently of one another. This led him to formulate his principle of independent assortment, in which he stated that any one of a pair of characteristics may combine with any one of another pair.

We can explain the results of Mendel's experiments in terms of what we know about alleles and the behaviour of chromosomes during meiosis. The pure-breeding plants, being diploid, will possess two identical alleles for each character. During gamete formation, meiosis occurs and the gametes will contain one allele for each character. In the F_1 generation, the genotype can only be **RrYy**, and when these plants are self-pollinated, there are four possible combinations in both the male and female gametes: **RY**, **Ry**, **rY** and **ry**. Because fertilisation is random, and any male gamete can fuse with any female gamete, there are 16 possible combinations, as shown in the Punnett square. We know there are four different phenotypes, but these are produced by nine different genotypes. These outcomes are summarised in Table 6.3.

In order to verify that there were nine different genotypes and that there were four possible combinations in the gametes, Mendel carried out further crosses. He grew the plants from the seeds of the F_2, allowed them to self-pollinate and collected all the progeny from this F_3 generation. He found that the F_2 plants which had wrinkled seeds and green cotyledons only gave rise to plants that produced wrinkled seeds and green cotyledons. With the plants that produced wrinkled seeds and yellow cotyledons in the F_2, he found one-third of them bred true, producing offspring with wrinkled seeds and yellow cotyledons, but two-thirds of them did not. They were pure-breeding for wrinkled seeds, but the cotyledon character appeared in the ratio 3 yellow : 1 green. He obtained similar results for those plants that had round seeds and green cotyledons, but in this case all the plants bred true for the cotyledon character and it was the seed form that varied. Mendel also carried out a test cross using the F_1 genotype and the pure-breeding parental type with wrinkled seeds and green cotyledons. The results of this test cross, which are illustrated in

Figure 6.13 Dihybrid inheritance of independent characters, illustrated by Mendel's cross between pea plants with round seeds and yellow cotyledons and plants with wrinkled seeds and green cotyledons. Following self-pollination of the F_1 plants, the ratio of phenotypes in the F_2 generation is 9 : 3 : 3 : 1.

Parents: RRYY   ×   rryy
round seeds / yellow cotyledons   wrinkled seeds / green cotyledons

Gametes: (RY)   (ry)

F_1: RrYy — round seeds, yellow cotyledons

F_1 allowed to self-pollinate: RrYy × RrYy

Gametes: (RY) (Ry) (rY) (ry)

	(RY)	(Ry)	(rY)	(ry)
(RY)	RRYY round, yellow	RRYy round, yellow	RrYY round, yellow	RrYy round, yellow
(Ry)	RRYy round, yellow	RRyy round, green	RrYy round, yellow	Rryy round, green
(rY)	RrYY round, yellow	RrYy round, yellow	rrYY wrinkled, yellow	rrYy wrinkled, yellow
(ry)	RrYy round, yellow	Rryy round, green	rrYy wrinkled, yellow	rryy wrinkled, green

F_2: 9 round, yellow: 3 round, green: 3 wrinkled, yellow: 1 wrinkled, green

Figure 6.14, provided evidence that four types of gametes were produced and that they were present in equal proportions.

As we have seen, in this dihybrid cross, two of the phenotypes in the F_2 were like the original parents and two showed new combinations of characters. This process by which new combinations of characters arise is known as **recombination** and the individuals that have the new combination of characters are called **recombinants**. Recombination is important as it can lead to genetic variation and accounts for many of the small differences between individuals in a natural population.

Dihybrid inheritance can be demonstrated in other organisms, and the cross illustrated in Figure 6.15 shows the inheritance of body colour and wing form in the fruit fly, *Drosophila melanogaster* (see Figure 6.2). *Drosophila* has been used extensively in genetics experiments because it is easy to keep under laboratory conditions, it has a short life cycle of 10 days and only four pairs of chromosomes.

Linkage

In the early 1900s, after the rediscovery of Mendel's work, many experimental dihybrid crosses were carried out in an attempt to verify Mendel's ratios, but not all of these gave the 9 : 3 : 3 : 1 ratio. One such cross was carried out by Bateson and Punnett, using sweet peas with different flower colour and pollen characteristics. They crossed sweet peas having purple flowers and long pollen

Table 6.3 *Phenotypes and genotypes in dihybrid inheritance*

Phenotype	Genotypes
round seeds, yellow cotyledons	RRYY RRYy RrYY RrYy
round seeds, green cotyledons	RRyy Rryy
wrinkled seeds, yellow cotyledons	rrYY rrYy
wrinkled seeds, green cotyledons	rryy

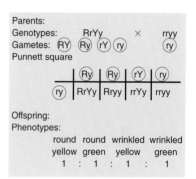

Figure 6.14 *Test cross in dihybrid inheritance, showing that four types of gametes are produced in equal proportions by the F_1 (heterozygous, RrYy) plants*

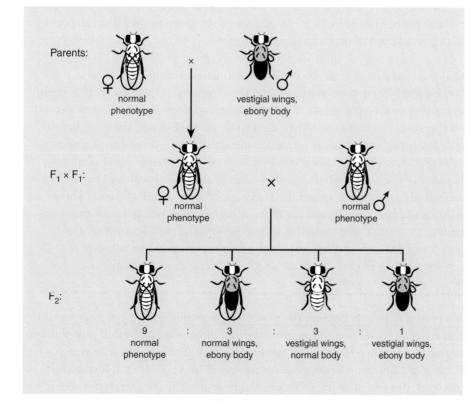

Figure 6.15 *Dihybrid cross in* Drosophila, *illustrating inheritance of wing form and body colour, and showing that the alleles for vestigial wings and ebony body are recessive*

GENES, ALLELES AND SOURCES OF NEW INHERITED VARIATION

Table 6.4 *Results of Bateson and Punnett's experiment*

Phenotype	Results	Expected results
purple flower, long pollen	296	240
purple flower, round pollen	19	80
red flower, long pollen	27	80
red flower, round pollen	85	27

with plants having red flowers and round pollen. The F_1 plants had purple flowers and long pollen, so Bateson and Punnett knew which were the dominant traits. The F_1 were allowed to self-pollinate and in the F_2 most of the offspring resembled the original parents, with a small number of recombinants. The results are shown in Table 6.4. Alongside Bateson and Punnett's figures are the numbers expected if the F_2 ratio was 9 : 3 : 3 : 1.

If we look at just flower colour, we can work out that there were 315 purple-flowered plants and 112 red-flowered plants, which is approximately a 3 : 1 ratio. Looking at the results overall, the ratio of the original parental types is much closer to a 3 : 1 ratio than it is to a 9 : 3 : 3 : 1. The results of this experiment, together with other evidence, led geneticists to suggest that some characters tend to be inherited together. In the light of our knowledge of the structure of chromosomes, it now seems obvious that there are large numbers of genes on each chromosome and that these genes will usually be inherited together. What seems surprising is that Mendel selected characters that are on separate chromosomes, especially as the haploid number of chromosomes in the garden pea is seven. Was he lucky, or did he select his characters carefully?

Genes located on the same chromosome are said to be **linked** and a **linkage group** consists of all the genes on a single chromosome. If the single chromosome does not determine the sex of the individual, then the genes on it show **autosomal linkage.** In Bateson and Punnett's experiment, the genes for flower colour and pollen shape are on the same chromosome and inherited together. This accounts for the large numbers of the original parental types in the F_2, but we can see that the linkage is not complete as there are small numbers of recombinants. These can be accounted for by the crossing over that can occur during prophase I of meiosis (Figure 6.16).

Complete linkage, where there is no crossing over during meiosis and consequently no recombinants among the offspring, is rare, but it does occur in some flowering plants, such as *Fritillaria*, and a few insects. Genes located on the same chromosome more often show partial linkage, because there is usually at least one chiasma formed somewhere within a bivalent. Crossing over produces two kinds of recombinants in approximately equal numbers, because of the way in which the non-sister chromatids break and rejoin, but it will not always occur between the two genes concerned. Figure 6.17 shows what can happen: in *(a)* and *(b)* there is no crossing over between the two genes, but in *(c)* there is. As this will occur in some cells and not others, and because it only involves two of the four chromatids, the number of recombinants is likely to be small and will always occur in fewer than 50 per cent of the progeny.

If an unusual ratio is obtained in a breeding experiment, it is necessary to do a test cross or back cross to discover whether linkage is involved. If independent segregation occurs, then the expected ratio in a test cross between the F_1 progeny and the homozygous recessive will be 1 : 1 : 1 : 1. If linkage is involved, then most of the offspring will resemble the parental types with a small number of each of the recombinants. The two parental types will occur in approximately equal numbers, as will the recombinants (Figure 6.18).

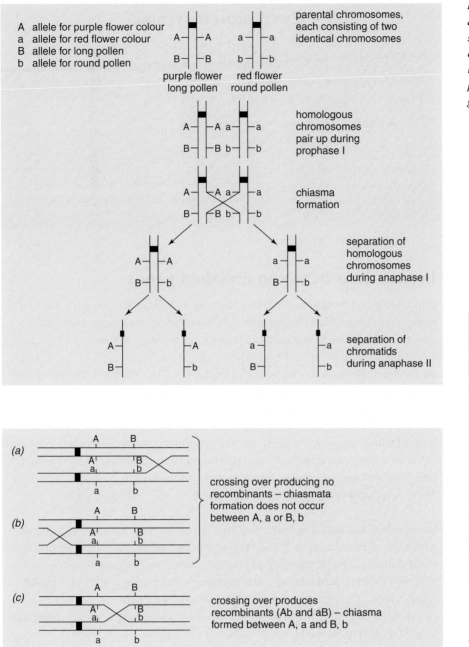

A   allele for purple flower colour
a   allele for red flower colour
B   allele for long pollen
b   allele for round pollen

purple flower long pollen   red flower round pollen

parental chromosomes, each consisting of two identical chromosomes

homologous chromosomes pair up during prophase I

chiasma formation

separation of homologous chromosomes during anaphase I

separation of chromatids during anaphase II

Figure 6.16 Explanation of Bateson and Punnett's results with sweet pea, showing how crossing over between chromatids during prophase I of meiosis can produce a small proportion of recombinants during gamete formation

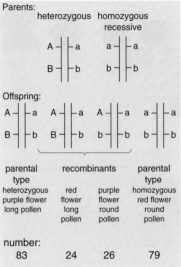

Parents:
heterozygous   homozygous recessive

Offspring:

parental type	recombinants		parental type
heterozygous purple flower long pollen	red flower long pollen	purple flower round pollen	homozygous red flower round pollen

number:

| 83 | 24 | 26 | 79 |

Figure 6.18 Results of test cross with sweet pea, confirming that the genes for flower colour and pollen shape are linked in sweet pea and that recombination has occurred to separate the alleles in the linkage group

(a)

(b)

crossing over producing no recombinants – chiasmata formation does not occur between A, a or B, b

(c)

crossing over produces recombinants (Ab and aB) – chiasma formed between A, a and B, b

Figure 6.17 Diagrams illustrating that the site of chiasma formation and subsequent crossing over between chromatids determines whether or not recombinants occur during gamete formation

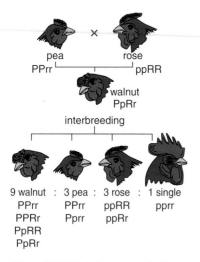

Figure 6.19 Variation in comb shape in poultry due to the interaction between unlinked genes affecting the same characteristic

Interactions between unlinked genes

In the examples of dihybrid crosses discussed so far in this chapter, each of the genes controls a different character, but there are many instances where a single character is influenced by two or more unlinked genes.

Bateson and Punnett worked out the interactions between two unlinked genes that are involved in the determination of comb shape in domestic poultry. There are a number of different forms of the comb and Bateson and Punnett crossed birds with 'pea' combs and birds with 'rose' combs. They knew that both of these types bred true, but the F_1 produced a completely different type of comb shape, called 'walnut'. Interbreeding of these walnut types yielded pea, rose and walnut combs, together with yet another type known as 'single'. The four phenotypes occurred in the ratio 9 : 3 : 3 : 1 (Figure 6.19).

Bateson and Punnett suggested that there were two genes involved: one in which the dominant allele **P** gave rise to the pea form and the other in which the dominant allele **R** gave rise to the rose form. So, the genotypes of the parents in their original cross must have been **PPrr** (pea comb) and **ppRR** (rose comb). The genotype of the F_1 was therefore **PpRr**, interacting to give the walnut comb. In the F_2, the genotype with no dominant alleles, **pprr**, has a single comb (Figure 6.20).

The ratios obtained in this cross are the same as Mendel obtained in his dihybrid crosses and indicate that the two genes are inherited independently. The walnut and single combs are completely new forms of the character, and are due to the interactions between the genes concerned and cannot be thought of as recombinants.

There are other examples of gene interaction where the ratios are different, but still seem to be multiples of $\frac{1}{16}$. In another investigation carried out by Bateson and Punnett, two pure-breeding white-flowered sweet pea plants produced offspring with purple flowers. These purple-flowered plants were self-pollinated and the resulting progeny consisted of plants with purple flowers, and plants with white flowers in the ratio 9 purple : 7 white. In this case, it was suggested

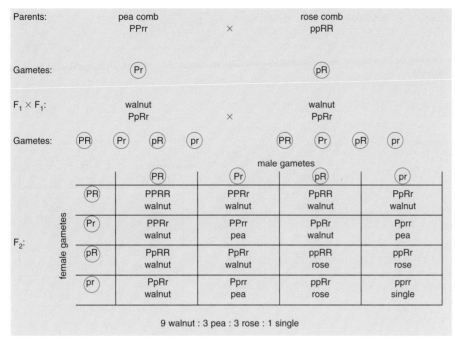

Figure 6.20 Genetic cross to show how all four comb phenotypes can occur in the F₂ offspring, following interbreeding of the heterozygous, walnut comb F₁ generation

that one of the genes (**C**) controls a colourless precursor of the pigment and the other gene (**P**) controls the conversion of this precursor to its purple form. In order for the purple colour to develop, the dominant allele of both genes must be present. The two white-flowered plants must be homozygous dominant for one of the genes in order for purple colour to show up in the offspring. The interaction is summarised in Figure 6.21 and the cross is illustrated in Figure 6.22.

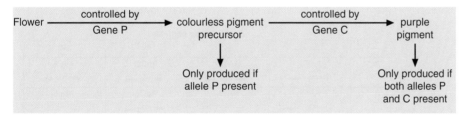

Figure 6.21 Simple diagram to show how genes for flower colour interact

These two genes controlling flower colour are said to be complementary, as they interact together to produce a character which is different from that given by either of them on their own.

Another example of this type of interaction is shown in the inheritance of coat colour in mice. Wild mice have a coat colour referred to as 'agouti', where the individual hairs are black with a yellow band. Agouti (**A**) is dominant to black coat colour (**a**), which is produced by black hairs. The difference is due to a single gene with two alleles. In order for the coat-colour pigment to be synthesised, at least one dominant allele of another, independently inherited, gene (**C**) is required. When individuals are homozygous recessive (**cc**) for this second gene, no pigment can be formed and the mice are albino. Figure 6.23 illustrates a cross between a pure-breeding black mouse (**aaCC**) and an albino (**AAcc**), which produces agouti mice (**AaCc**) in the F₁. When F₁ individuals are

> **QUESTION**
>
> Using the information in Figure 6.21, construct a table of phenotypes with their genotypes for flower colour in sweet pea plants.

Figure 6.22 Genetic cross to illustrate complementary interaction between two genes affecting expression of flower colour in sweet peas, showing that if one of the genes is homozygous recessive the purple colour is not expressed

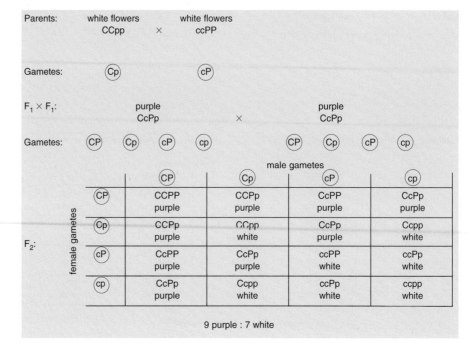

Figure 6.23 Inheritance of coat colour in mice by complementary interaction of genes, showing how the gene for hair pigment (C) masks the effect of the gene for coat colour (A) when it is homozygous recessive (cc)

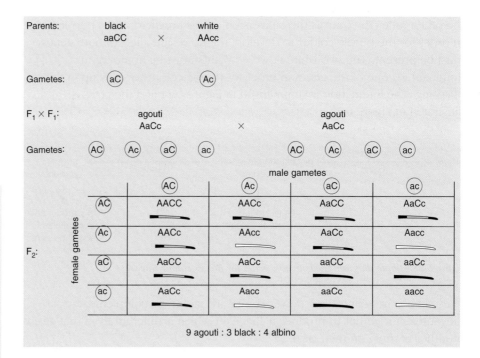

ADDITIONAL MATERIAL

Epistasis

The inheritance both of flower colour in the sweet peas and of coat colour in mice are examples of epistasis, where one gene hides the expression of another gene. In the sweet peas, where a dominant allele of both genes is required to produce flower colour, the homozygous recessive condition of either gene is epistatic to the expression of the other gene. In the mice, the gene for pigment development (**C**) is epistatic to the gene for coat colour when it is homozygous recessive (**cc**).

interbred, the resulting offspring include agouti, black and albino mice in the ratio 9 agouti : 3 black : 4 albino.

Sex determination in humans

Sex is genetically determined and matings between males and females of a species always result in two sexes among the offspring. The two sexes occur in approximately equal numbers, the same ratio as would be expected in a monohybrid test cross. Examination of the chromosomes of many mammals and insects indicates that sex determination is associated with the inheritance of a pair of particular chromosomes, designated the sex chromosomes. There are two kinds of sex chromosomes, called the X and the Y (see Figure 7.4 page 102). The rest of the chromosomes, apart from the X and the Y, are known as the autosomal chromosomes, or autosomes. The X chromosome is usually larger than the Y, but they pair because they have regions that are homologous. In the non-homologous, or non-pairing, region of the X chromosome there are genes that are not present on the Y chromosome. In many species, the Y chromosome carries very few genes and is often referred to as 'genetically empty'. It is important to realise that, although the sex chromosomes determine the sex of an individual, they do not carry all the genes responsible for the development of the sexual characteristics.

In humans, the female is referred to as the homogametic sex, as she possesses two X chromosomes (XX), and the male is the heterogametic sex, having one X chromosome and one Y chromosome (XY). The sex chromosomes pair during prophase I of meiosis. Like other paired chromosomes they are separated at anaphase I and chromatid separation occurs at anaphase II. Females can only produce eggs that contain a single X chromosome, but males produce sperm with either an X chromosome or a Y chromosome. Due to the events of meiosis, 50 per cent of the sperm carry an X and 50 per cent carry a Y. At fertilisation, there is an equal probability of an egg fusing with a sperm carrying an X as with a sperm carrying a Y. The sex of the offspring is thus determined by the male (Figure 6.24).

In the early stages of development of the human embryo, there are no external signs to indicate its sex, since the reproductive organs are still capable of developing into ovaries or testes. The presence of the Y chromosome triggers the development of testes, probably through the production of higher levels of male hormones. Absence of the Y chromosome results in the formation of ovaries.

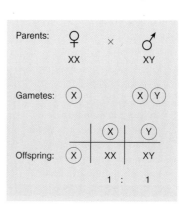

Figure 6.24 Sex determination by the X and Y chromosomes, showing how the 1 : 1 male : female ratio is achieved and how the male determines the sex of the offspring

> ### DEFINITION
> A **mutation** is a change in the structure or the amount of DNA in an organism.

Sources of new inherited variation

Mutation

During the process of replication, DNA is normally copied exactly so that the genetic material remains the same from generation to generation. However, occasionally a **mutation** occurs, resulting in a change in the structure or the amount of DNA present in an organism. Mutations may occur in the DNA of somatic (body) cells, but if a mutation occurs in cells during the formation of gametes, the mutation will be passed on to the next generation. Mutations occurring in the formation of gametes are referred to as germ-line mutations and are an important source of genetic variation within a population.

Mutation may occur spontaneously as a result, for example, of a mistake occurring during DNA replication. The frequency with which one allele mutates to another is known as the mutation rate. As an example, the mutation rate for purple maize grains is 1×10^{-5}. This means that out of 10^5 alleles for white grains, only one (that is, 1 in 100 000) mutates to an allele for the purple grain every generation (Figure 6.25). Altered DNA molecules replicate the changed nucleotide base sequences so that mutant genes are passed on to successive

Figure 6.25 Sweet corn (maize), showing production of purple and white seeds in a single cob

generations. This spontaneous mutation rate is essential for providing new variation necessary for survival in a changing environment; in other words, it provides the raw material for evolutionary change.

The mutation rate may be greatly increased by exposure to environmental agents, including certain chemicals and radiation. Such agents are referred to as **mutagens**.

Mutagens and point mutations

Mutagens are physical or chemical agents that greatly increase the mutation rate. Mutagens include ultraviolet light and X-rays, which damage DNA in various ways, and certain chemical substances, which alter DNA by adding or deleting one or more bases in a sequence. Chemical mutagens include nitrous acid and 5-bromouracil. Nitrous acid acts directly on nucleic acids, and alters the genetic code by converting one base into another. For example, cytosine is converted into uracil, which bonds to adenine instead of guanine. Such changes in the sequence of bases are referred to as **point mutations** and include **deletion**, **insertion** and **substitution** mutations. Figure 6.26 illustrates these types of mutation.

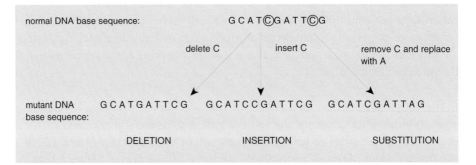

Figure 6.26 Mechanisms by which deletion, insertion and substitution mutations can arise

What are the consequences of these changes in the base sequence?

You will recall that, during protein synthesis, the sequence of bases in DNA is transcribed into a complementary sequence of bases in messenger RNA (mRNA), which is then translated into a sequence of amino acids. Each sequence of three bases is a code for a specific amino acid, so that if the code is altered, this may result in an incorrect sequence of amino acids. Consider the effect of base deletion, for example, on the sequence of amino acids (codons are separated to make the changes clearer):

normal DNA sequence:
GAC TTC AGT CTA

complementary mRNA sequence:
CUG AAG UCA GAU

amino acid sequence:
Leu – Lys – Ser – Asp

Now suppose that the first C in the normal DNA sequence is deleted:

mutant DNA sequence:
GAT TCA GTC TA

complementary mRNA sequence:
CUA AGU CAG AU

amino acid sequence:
Leu – Ser – Glu–

The effect on the sequence of amino acids from this small change in the DNA bases is profound. Every amino acid after the change might be replaced by a different amino acid. Base deletion has resulted in a shift of the 'reading frame' known as a frame shift, so that all subsequent codons are altered. Since the order of amino acids is responsible for the final shape and functional properties of the completed protein, it is not surprising that mutation can result in the synthesis of a totally different protein.

We can illustrate the effect of point mutation by reference to a specific example. **Sickle-cell anaemia** (see page 82) arises because of a substitution mutation in a single codon, which specifies valine instead of glutamic acid in the amino acid sequence of the β-chain in the haemoglobin molecule. This is illustrated in Figure 6.27. The affected codons only are shown.

QUESTION
Suggest what would be the likely effect of a deletion mutation followed by an addition mutation in a different part of the same gene.

Normal haemoglobin (haemoglobin A)

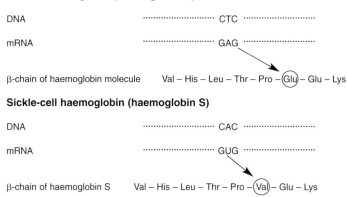

Figure 6.27 Effect of substitution mutation on expression of gene for normal haemoglobin (HbA) β-chains, leading to synthesis of sickle-cell haemoglobin (HbS)

This change has a profound effect on the properties of haemoglobin S, which can result in changes in the shape of the red blood cells. Such cells have a shorter life span than normal red blood cells and therefore sufferers are likely to become anaemic. Sickle-cell anaemia is a serious condition, which is often fatal before middle age.

Chromosome mutations

Chromosome mutations can arise in two major ways: either by an alteration in the structure of a chromosome or by an alteration in the number of chromosomes. Changes in structure include **deletion**, where part of a chromosome breaks off and is lost when the cell divides, and **translocation**, where a fragment of one chromosome attaches to another, non-homologous chromosome. One type of translocation, known as reciprocal translocation,

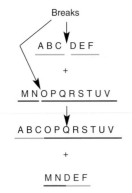

Figure 6.28 Reciprocal translocation of segments between two adjacent chromosomes

---

QUESTION

Distinguish between the terms polysomy and polyploidy.

---

involves a two-way exchange of chromosomal segments. This is illustrated in Figure 6.28. The letters are used to indicate regions of each chromosome.

Changes in the number of chromosomes can vary from the duplication or loss of one chromosome (**polysomy**) to duplication of every chromosome, so that there are three or more complete sets of chromosomes (**polyploidy**).

Both changes in the structure and in the numbers of chromosomes have consequences other than their immediate effects on the chromosomes. Individuals who are heterozygous for chromosomes with different structures frequently have reduced fertility, and individuals with altered numbers of chromosomes may be non-viable or sterile.

Translocation

A translocation is the movement of one part of a chromosome to another, non-homologous chromosome. Translocations can cause several human diseases. For example, about 5 per cent of individuals with **Down's syndrome** have one parent who is heterozygous for a translocation mutation. This arises because chromosome number 21 attaches to chromosome number 14, forming a translocation heterozyte. A gamete may be produced containing the translocated 14 with the attached 21, plus a normal 21. If this gamete is fertilised, three chromosomes number 21 (trisomy 21) are present in the zygote, which results in Down's syndrome. Fertilisation of the gamete containing no chromosome number 21 results in a zygote with only one 21 chromosome (monosomy 21), which is a lethal genotype. Overall, approximately one-third of the live births arising from this translocation heterozygote will be expected to have Down's syndrome. However, the proportion is less than this, because some Down's individuals do not survive gestation (Figure 6.29).

Figure 6.29 Mechanism by which individuals with Down's syndrome result from a translocation mutation (chromosome 14 is shown in blue and chromosome 21 in red)

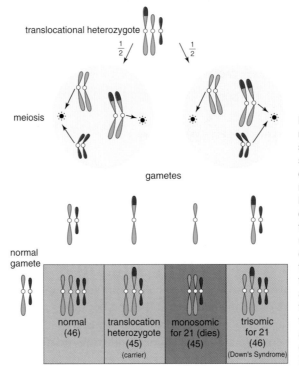

Half of the time, the heterozygote may produce either the normal set or a balanced translocated set of chromosomes. In this case the offspring are either normal or translocation heterozygotes, with 45 chromosomes. The other half of the time, meiosis produces unbalanced chromosomes, either a 14 without the translocated 21 segment, or a translocated 14 with the attached 21 plus a normal 21. In the second case, fertilisation with a normal gamete produces a zygote with three 21 chromosomes, resulting in Down's syndrome.

Polysomy

Polysomy is the general term for the condition in which the number of a particular chromosome is not diploid. Sometimes during meiosis, a pair of homologous chromosomes will fail to separate properly so that one daughter cell receives both chromosomes and one daughter cell is left with no chromosome from that pair. This failure of homologous chromosomes to separate is termed **non-disjunction**. When these gametes are fertilised by a normal gamete, the zygote will either have one extra chromosome, that is, $2n + 1$, which is termed trisomy, or it will have one missing chromosome, $2n - 1$, termed monosomy. Non-disjunction is most common during meiosis I, but it can also occur during meiosis II.

One example of non-disjunction leading to trisomy is Down's syndrome. Although, as described previously, about 5 per cent of cases are a result of a translocation mutation, in about 95 per cent of cases it results from non-disjunction of chromosome number 21 during meiosis. This is shown in Figure 6.30, where only chromosome number 21 is shown.

Down's syndrome is characterised by mental retardation, distinctive palm prints and a characteristic facial appearance (Figure 6.31). The incidence of Down's syndrome increases with the age of the mother. For example, the incidence is about 1 in 2300 for mothers aged 20, but rises to about 1 in 40 for mothers aged 45. The exact reason for this is unknown.

Polyploidy

Organisms with three or more complete sets of chromosomes are known as **polyploids**. Polyploidy is relatively common in plants, but rare in animals, occurring only, for example, in certain beetles, earthworms and fish. Many important crop plants are polyploids: potatoes are tetraploid and bread wheat is hexaploid.

There are two major types of polyploids: **autopolyploids**, which receive all their chromosomes from the same species, and **allopolyploids**, which receive their chromosomes from different species. Polyploids can arise naturally when a cell undergoes abnormal meiosis and all the chromosomes go to one pole. This will result in a gamete having the diploid number of chromosomes. In most situations, this diploid gamete would combine with a haploid gamete, resulting in a triploid zygote.

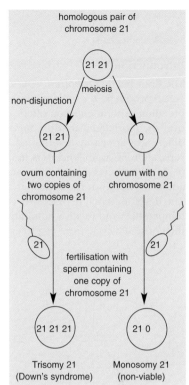

Figure 6.30 Non-disjunction of chromosome 21, leading to Down's syndrome

Figure 6.31 Child with Down's syndrome

GENES, ALLELES AND SOURCES OF NEW INHERITED VARIATION

A breeding experiment using *Tribolium*

Introduction

The flour beetle (*Tribolium confusum*) is a pest of stored cereal products. *T. confusum* is a useful organism for breeding experiments and offers a number of advantages over the more familiar *Drosophila*. For example, sex differences are easily determined at the pupal stage, so separation of virgin females is more straightforward (Figure 6.32). The insects do not climb smooth surfaces and the adults do not fly at room temperature.

The developmental periods vary considerably according to temperature and relative humidity (RH). For *T. confusum*, the entire life cycle from egg to egg takes 5 to 6 weeks at

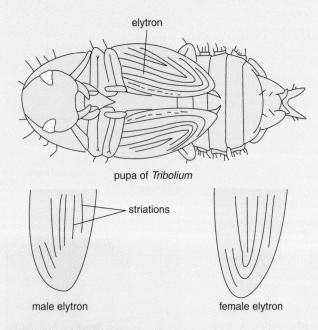

elytron

pupa of *Tribolium*

striations

male elytron   female elytron

Figure 6.32 Identification of male and female pupae of Tribolium confusum *according to markings (striations) on their elytra*

25 °C and 70 % RH. Insects can be reared in glass tubes (10 cm × 2.5 cm approx.) one-third filled with either wholemeal flour or wheat meal. To maintain constant conditions, these culture tubes should be kept in a desiccator, or other suitable container, within an incubator at 25 °C. A humidity of 70 % RH can be maintained using a saturated solution of sodium nitrate, placed in the desiccator. The lid should be removed for 30 seconds each day to air the cultures.

When setting up genetic crosses, it is essential to start with virgin females, because females can continue to lay eggs fertilised by stored sperm for some time after the initial mating. The determination of sex of *Tribolium* is best undertaken at the pupal stage. In *T. confusum*, the elytral (anterior wing) striations in the male do not reach the apex, whereas in the female, some of the striations meet at the apex.

Materials

- Stock cultures of *Tribolium confusum*, such as wild-type and pearl-eyed
- Desiccator or other suitable container
- Saturated sodium nitrate solution
- Incubator at 25 °C
- Culture tubes, 10 cm × 2.5 cm approx.
- Wholemeal flour or wheat meal
- Sieve
- Small paintbrush to manipulate insects
- Binocular microscope or hand lens

TOXIC
sodium
nitrate

Method

1. Sieve the stock cultures, sex and segregate male and female pupae. Incubate until adult. Alternatively, crosses may be made up starting with pupae.
2. Set up crosses, such as one female wild-type and two pearl-eyed males, in tubes of food medium.
3. Incubate at 25 °C and 70 % RH.
4. After 40 to 45 days, remove the adults. Then, once or twice a week, remove all emerged F_1 adults and set up crosses for the F_2.
5. Incubate the F_2 crosses and proceed as in step 4. Collect and score the F_2 adults until the whole generation has emerged.

Results and discussion

1. What phenotypic ratio would you expect in your F_2?
2. Calculate the expected numbers of each phenotype in your F_2.
3. Apply a chi-squared test to investigate whether or not your results differ significantly from the expected results.

Further work

1. Set up crosses other than that suggested.
2. Investigate the effect of temperature on the developmental periods of *Tribolium*.

Genetic counselling

Developments in molecular biology and medical research are giving us a greater understanding of a range of disorders or diseases which have a genetic cause and can therefore be inherited. **Genetic counsellors** try to help individuals to find out about the risk to themselves or to their children of suffering from a genetic disorder. When the possible risk has been established, or a genetic condition confirmed, the counselling process will try to help the person or family to decide what action to take, to learn what treatment may be available or sometimes just to come to terms with a difficult and distressing situation. The investigation and counselling process is likely to involve doctors, geneticists, biochemists and other analysts, as well as social workers.

Causes of genetic disorders

Genetic disorders may result from defects or abnormalities in the **chromosomes** or within the individual **genes**. The estimated prevalence of disorders caused directly by these genetic abnormalities is in the region of 20 per 1000 population. Other conditions, such as coronary heart disease, diabetes and cancer, have strong genetic links. Chromosome abnormalities may be due to changes in chromosome number or to structural defects, such as deletion or translocation. Abnormalities may be recognised from the **karyotype** of an affected individual. Many gross defects probably do not survive and may be a cause of early spontaneous abortion of a fetus. Defects in a single gene follow a Mendelian pattern of inheritance, expressed either in the **autosomes** or in the **sex (X or Y) chromosomes**, and may affect the dominant or the recessive allele. As a result of the Human Genome Project (a huge international research programme in which all the human genes are being mapped), the positions of many of the genes on a particular chromosome are known. Some examples of genetic disorders, their causes and descriptions of how they may affect people are given in Table 7.1.

Family history and risk of inheriting genetic disorders

A couple who know of an inherited disease in a near relative may seek advice through genetic counselling before deciding to start a family. With an understanding of Mendelian inheritance it may be possible to trace the family history through pedigree analysis and to work out the risk of an inherited disorder passing to their children (see Figure 7.1). Estimation of risk may alter depending on the age at which the onset of the disorder usually occurs. It may also be important to identify **carriers** – such people carry the affected allele but show no symptoms of the disorder. The risk of chromosome abnormalities increases with age of the mother so, for example, as we have seen, the risk of Down's syndrome in babies born to mothers over the age of 35 years is considerably higher than that for younger women.

GENETIC COUNSELLING

Table 7.1 *Genetic disorders – causes and descriptions in affected person*

Example	Cause and description of affected person
Chromosome abnormalities	
numerical defects	
Down's syndrome	• trisomy (2*n* + 1) of chromosome 21. Characteristic flat face with slanting eyes, broad hands, short in height, varying degrees of mental retardation, likelihood of other conditions such as heart defects and deafness.
Turner's syndrome	• monosomy (2*n* – 1) of X chromosome (XO). Females, but infertile as no ovaries, short in height and may have webbed appearance of the neck.
Klinefelter's syndrome	• 47 chromosomes (XXY). Males, but small testes and usually infertile (no sperm production). Tall and thin, possibly some development of breasts, sometimes mild mental retardation.
structural defects	
'cri du chat' syndrome	• deletion of short arm of chromosome 5. Mental retardation, multiple physical abnormalities, characteristic cry (like a cat).
fragile X syndrome	• fragile site on chromosome X. Mental retardation, abnormal head and face including high forehead.
Gene abnormalities	
autosomal dominant disorders	
achondroplasia	• dwarfism due to a defect in cartilage and bones so they fail to grow to the correct size. Intelligence, head and body size are all normal.
Huntington's disease (chromosome 4)	• involuntary jerky movements of the head, progressing into dementia. Starts to appear in early middle age.
adult polycystic kidney disease (chromosome 16)	• cysts in the kidney, may lead to kidney failure.
autosomal recessive disorders	
cystic fibrosis (chromosome 7)	• lacks the protein which allows transport of chloride ions across cell membranes. Production of thick mucus likely to affect pancreas, bronchi, sweat glands, etc. Frequent respiratory infections.
galactosaemia (chromosome 9)	• unable to utilise galactose (derived from lactose, but normally converted to glucose in liver). If untreated, babies become mentally retarded; they develop normally if fed on a galactose-free diet.
phenylketonuria (chromosome 12)	• has excess of the amino acid phenylalanine which damages the nervous system leading to mental retardation. If detected early, babies are fed on a diet free of protein containing phenylalanine.
sickle-cell anaemia (chromosome 11)	• produces abnormal sickle-cell haemoglobin (HbS), leading to distortion of the red blood cells and may result in anaemia. The defect is due to substitution of valine for glutamic acid in one chain in the haemoglobin molecule.
X-linked recessive disorders	
colour blindness	• most commonly as a failure to distinguish between reds and greens.
Duchenne muscular dystrophy	• weakness and wasting of muscles, especially of the back and pelvic girdle. Starts before the age of 4. Can be relieved but not cured by physiotherapy and orthopaedic treatment.
haemophilia	• blood clots very slowly because of lack of coagulating factors, known as Factor VIII and Factor IX. Person may suffer prolonged bleeding after wounding or spontaneous internal bleeding. Treatment by blood transfusion or supplying missing factor(s).

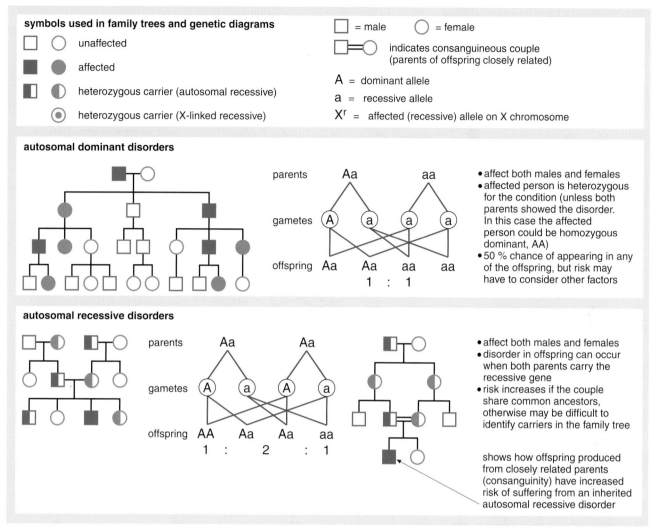

Figure 7.1 Family pedigrees, showing inheritance of autosomal dominant and autosomal recessive disorders

Detection of fetal abnormalities

Prenatal tests may be undertaken to allow detection of abnormalities in the fetus. Procedures commonly used include **ultrasound scanning**, **amniocentesis** and **chorionic villus sampling** (Figure 7.3). When some fetal material has been obtained, tests can be carried out to determine whether or not there are chromosomal or genetic defects and to identify them as far as possible. A **karyotype** is obtained by treating extracted cells so that the chromosomes become visible and they are then photographed. The chromosomes are cut out then arranged and numbered (from 1 to 22) in order of size, starting with the largest, and followed by the sex chromosomes (X and Y) (Figure 7.4).

DNA analysis is now becoming a standard procedure for locating particular gene defects and for identifying carriers (Figure 7.5). The pattern of bases in a DNA molecule can be achieved using the technique known as **DNA profiling**.

GENETIC COUNSELLING

ADDITIONAL MATERIAL

Family pedigrees showing sex-linked disorders

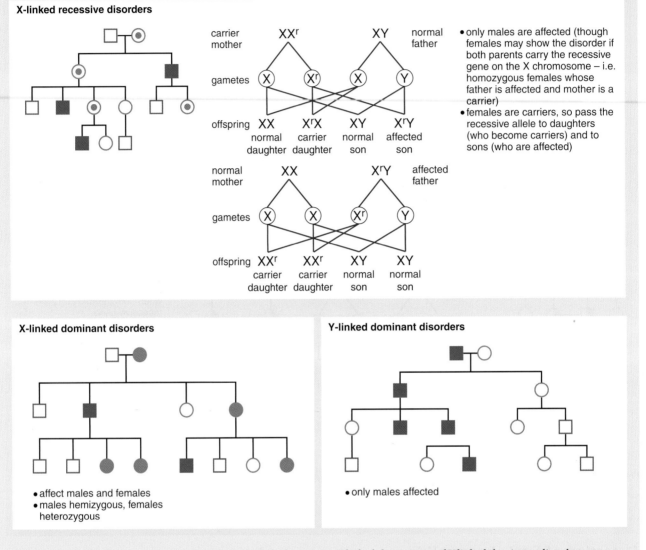

- only males are affected (though females may show the disorder if both parents carry the recessive gene on the X chromosome – i.e. homozygous females whose father is affected and mother is a carrier)
- females are carriers, so pass the recessive allele to daughters (who become carriers) and to sons (who are affected)

Figure 7.2 Family pedigrees, showing inheritance of X-linked recessive, X-linked dominant and Y-linked dominant disorders

The DNA is extracted from the fetal material, then cut into lengths using enzymes. The enzymes which cut DNA are known as **restriction enzymes**, and these enzymes recognise particular sequences so can be chosen to cut the DNA molecule in a specific place. The fragments of DNA are loaded onto an agarose gel, then separated by electrophoresis. The resulting patterns of banding of the DNA fragments can be revealed by treatment with radioactive labels or dyes. Techniques are also available to work out the actual sequence of bases along a length of DNA.

Gene probes can be used to identify particular sequences in the DNA molecule. A probe is a short length of nucleotides with unique base sequences which binds to and hybridises with particular portions of a complementary

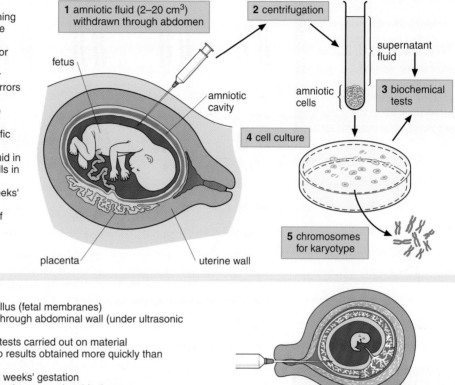

ultrasound scanning
- uses high frequency sound waves to build up picture on TV screen of features within the body
- confirms viable pregnancy
- locates placenta
- monitors fetal growth
- detects major deformities

amniocentesis
- sample of amniotic fluid containing fetal cells withdrawn through the abdominal wall
- amniotic cells can be cultured for further tests
- biochemical analysis on fluid or cultured cells for diagnosis of errors of metabolism
- cells can yield karyotype to see chromosome pattern
- analysis of DNA to locate specific gene defects
- results can be obtained from fluid in about 1 week, from cultured cells in about 3 to 4 weeks
- performed at about 15 to 16 weeks' gestation
- reliable and safe with low risk of increasing miscarriage

1 amniotic fluid (2–20 cm^3) withdrawn through abdomen

2 centrifugation

supernatant fluid

fetus

amniotic cavity

amniotic cells

3 biochemical tests

4 cell culture

5 chromosomes for karyotype

placenta

uterine wall

chorionic villus sampling
- material taken from chorionic villus (fetal membranes)
- removed through the cervix *or* through abdominal wall (under ultrasonic guidance)
- biochemical and chromosomal tests carried out on material
- culturing of cells not needed (so results obtained more quickly than amniocentesis)
- performed at between 9 and 12 weeks' gestation
- risk of miscarriage slightly higher than normal at this time

Figure 7.3 Prenatal testing of fetus: outline of ultrasound scanning, amniocentesis and chorionic villus sampling

molecule of DNA. In detection of genetic defects, the probe is chosen to correspond to and recognise regions where the faulty base sequence is suspected. Probes are now used successfully to detect inherited conditions such as Huntington's disease, cystic fibrosis, Duchenne muscular dystrophy and phenylketonuria.

Social, ethical and legal considerations

As research developments in reproductive biology continue, individuals and couples have more and more control over their own reproduction: if and when to have children, means of overcoming infertility, understanding of the risk of producing children with genetic disorders, knowledge of whether a fetus has certain genetic abnormalities. With the information available, individuals need to make decisions, as to whether to start a family or to have further children, whether to terminate a pregnancy in which genetic defects have been

GENETIC COUNSELLING

QUESTIONS

Find out about DNA technology ... and how it helps detect genetic defects.

- Why is the polymerase chain reaction (PCR) important in DNA analysis?
- How is the DNA sample broken into fragments?
- What technique is used to separate the DNA fragments?
- How is the gene defect identified?

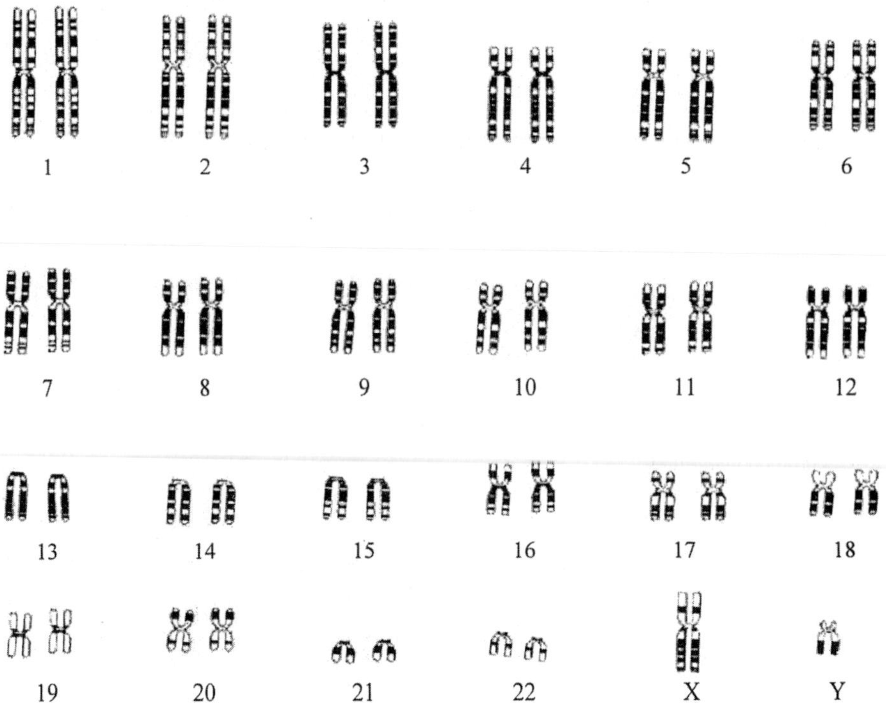

Figure 7.4 A human karyotype. Metaphase chromosomes are photographed, cut out and arranged according to size. There are 22 pairs of autosomes (non-sex chromosomes) and a pair of sex chromosomes, labelled X and Y.

Figure 7.5 With advances in DNA technology, many genes on the human chromosomes have now been mapped, including those for certain genetic disorders. Analysis of selected portions of DNA can lead to detection of mutant alleles, indicating specific genetic disorders. The photo shows autoradiagrams of DNA fragments, separated by electrophoresis, being inspected on a lightbox.

confirmed or take advantage of the treatment available to help the affected new-born baby live and grow through to adulthood. The decisions often need to be made within a framework of informed advice, with guidance from medical experts and consideration of the legal requirements as well as personal and often religious beliefs. For the people involved in a reproductive dilemma, to come to a decision may be emotional and stressful and it may take time and sympathetic understanding to come to terms with the situation.

For those faced with risk or reality of genetic defects in their children, genetic counselling aims to help individuals go through all aspects of the situation. Studies of family history allow an assessment of risk and may suggest priority is given to prenatal testing. Genetic screening before pregnancy can detect unaffected carriers and be offered to the individuals concerned as well as to close relatives. Genetic screening by tests during pregnancy may be offered to those most at risk and, as more tests become available through DNA analysis, it is likely that prenatal screening will be used to diagnose a wide range of possible defects. It is essential that the tests are reliable and that the results are accurate: to obtain either false negative or false positive information would be extremely distressing. It is important for the tests to be carried out early in pregnancy, partly to relieve anxiety for the pregnant mother and her partner but also so that there is opportunity for termination of pregnancy before the 13th week, if possible, if it is decided to take this course of action. In this respect, chorionic villus sampling is an improvement over amniocentesis, though it carries an increased risk of spontaneous abortion.

Termination of pregnancy by abortion may be **spontaneous** (also known as miscarriage) or **induced** artifically. It is estimated that about 15 to 20 per cent of conceptions end in spontaneous abortion, often within the first 3 months of pregnancy, and probably half of these carry chromosomal abnormalities. Induced abortion may be carried out surgically or by administering drugs, such as prostaglandins, which cause uterine contractions so that the immature fetus is expelled. When undertaken with good medical care, induced abortions are generally safe. However, many illegal abortions, particularly those used as a form of contraception for unwanted pregnancies, are performed in unhygienic conditions which can be dangerous for the mother.

In the UK, induced abortion can only be performed within the terms of the Abortion Act (1967) and later legal Abortion Regulations. The requirements are for two doctors to agree to termination of the pregnancy, for reasons stated in the Act, and for it to be carried out in an approved hospital or clinic. Even so, some people, for personal or religious reasons, will not agree to have an abortion and there is continuing active debate as to the stage at which the fetus has the 'rights' of a person. Some people believe it is at the moment of fertilisation, others accept abortion right up to the moment of delivery of the baby if it is in the interest of the physical or mental health of the mother, or if there is a significant risk that if the baby is born it will suffer from a serious physical or mental handicap. In some countries (for example, the Republic of Ireland and Portugal), therapeutic abortion is illegal at all stages whereas other countries have more liberal pro-abortion policies. In the UK, the 1967 Abortion Act allowed abortion (for specified reasons) up to 28 weeks of gestation, though more recently this limit has been relaxed in certain cases. Inevitably the debate will continue and one factor to be considered is that babies born as early as 23 weeks can now survive. This emphasises the need for accurate diagnosis of potential defects to be made early in pregnancy and for informed counselling to help the couple come to a decision which is consistent with their personal beliefs and the legal constraints.

Whilst medical technology allows manipulation of reproduction and inevitably raises ethical dilemmas, it also offers treatment or cures for some of the genetic disorders which can be detected by prenatal screening. Galactosaemia and phenylketonuria (PKU) are conditions that can both be controlled by careful selection of diet. For sufferers from galactosaemia, it is essential for the baby to receive a galactose-free or lactose-free diet soon after birth, and probably for life, so avoiding development of permanent mental retardation and other symptoms. Similarly, those affected by PKU can be treated by limiting the intake in the diet of the amino acid phenylalanine. For some conditions, it may be possible to supply the missing gene product and examples here include treatment for haemophilia, growth hormone deficiency and insulin deficiency.

For many genetic disorders for which there is no permanent cure or relief, there is a network of support groups for families, and such groups often offer advice or information which may help the family come to terms with the condition and allow the best quality of life available for the affected person. Details of the support network may be obtained through: *Contact a family, 16 Strutton Ground, London SW1P 2HP.*

GENETIC COUNSELLING

A further prospect of cure for people with genetic defects, and one which is undergoing active (and controversial) research, is the use of **gene therapy**. This utilises gene technology to incorporate a portion of DNA carrying the required correct base sequence into the genome of the affected person so that the genetic defect is effectively repaired. In 1997 clinical trials were in progress for gene therapy for cystic fibrosis patients. Whilst this approach has considerable potential for offering hope to sufferers, at the same time it raises ethical concerns over manipulation of the human genome and how far medical research can go. As with research on human embryos and the possibilities in the future for cloning of human beings, we need to give very careful thought concerning the controls to be set up to ensure the techniques are used in a responsible manner. There is no doubt that the technology is there, or will be in the forseeable future, but how it is used remains a personal as well as a legal decision, and one that needs debate on an international scale to provide appropriate safeguards for our future societies.

We need to keep in perspective the research advances and future potential of reproductive and DNA technologies. At one time, Edward Jenner was a brave man to experiment with the use of cowpox vaccine to provide immunity against the infectious disease of smallpox. Despite initial and continuing controversy, Jenner's trials resulted in development of vaccines which have led to world-wide eradication of smallpox. We now accept heart by-pass operations and organ transplant surgery as realistic ways of prolonging and improving the quality of life for sufferers of certain medical conditions. Yet at the time of the first attempts, the pioneers required considerable faith in their research developments and had to withstand controversial debate before they achieved a degree of success. However, there will always be some people who cannot accept interference with natural events in the human body and disagree with any form of artificial contraception, termination of pregnancy or even diagnosis of genetic defects. Their views must be respected. Others are prepared to move forward with the frontiers of scientific knowledge, but application of biological technology must be in the context of informed debate and responsible decision making.

Preparation of a karyotype

Introduction

A **karyotype** is a photographic or pictorial representation of all the chromosomes present in a somatic (body) cell of an individual. To prepare a karyotype, cells such as lymphocytes are cultured so that they are actively dividing, then the cells are broken open and the chromosomes spread onto a microscope slide. Various chemical methods are used to identify certain regions of the chromosomes by staining them so that they form darkly stained **bands**. The chromosomes are then photographed under a microscope and arranged systematically to produce the karyotype. In humans, a system is used for identifying and arranging the chromosomes based on:

- the overall size of the chromosome
- the position of the centromere
- the banding pattern.

The centromere appears as a constriction where the chromosome appears to be 'pinched'. If the centromere is near the middle of the chromosome, the chromosome is said to be **metacentric**; if the centromere is towards one end, the chromosome is **acrocentric**. The centromere therefore divides the chromosome in two arms; an acrocentric chromosome has one short arm and one long arm. By convention, the short arm is designated **p** and the long arm **q**.

In the preparation of a karyotype, the autosomes (chromosomes other than the sex chromosomes) are numbered from 1 to 22 on the basis of their overall length; the X and the Y chromosomes are identified separately. Human chromosomes are also classified into the following groups (referred to as the Denver System):

- Group A   chromosome numbers 1 to 3
- Group B   chromosome numbers 4 and 5
- Group C   chromosome numbers 6 to 12
- Group D   chromosome numbers 13 to 15
- Group E   chromosome numbers 16 to 18
- Group F   chromosome numbers 19 and 20
- Group G   chromosome numbers 21 and 22.

It is interesting to note that the correct diploid number of human chromosomes (46) was confirmed in 1956; for many years previously the number was incorrectly thought to be 48.

Materials

- Human chromosome analysis set (available from Philip Harris Education)
- Scissors

Method

1 Cut out each chromosome from the print supplied.
2 Using the banding pattern for reference, prepare your own karyotype for a male and a female individual.
3 Prepare karyotypes for individuals with Down's syndrome and Klinefelter's syndrome.

Results and discussion

1 Label your karyotypes fully, including the numbers of each pair of autosomes and the X and Y chromosomes.
2 Write an account to explain the chromosomal events which result in Down's syndrome and Klinefelter's syndrome.

Further work

Find out about each of the following techniques and explain how they are used in chromosome studies:

- amniocentesis
- chorionic villus sampling
- fluorescence *in situ* hybridisation (FISH).

Environmental change and evolution

The age of the Earth is estimated to be 4.6×10^9 years and the first forms of living organisms appeared about 4×10^9 years ago. These included bacteria-like and algae-like microorganisms. Now there are millions of different kinds of organisms on Earth, ranging from bacteria and single-celled fungi to complex flowering plants, insects and mammals.

Over this enormous period of time, organisms have changed dramatically. Whole groups, such as **trilobites** (early arthropods, Figure 8.1) and the **dinosaurs**, flourished and then became extinct. In general, as time passed, the structure of newly evolved organisms became progressively more complex. For example, marine algae appeared about 590 million years ago (mya), the first land plants about 400 mya and flowering plants first appeared in the fossil record 145 mya. Humans and their immediate ancestors have been on Earth for only about the last 2 million years.

Figure 8.1 A fossil trilobite, Olenoides

The process by which these changes have occurred over geological time is termed **evolution** and the branch of science which is concerned with the study of fossils is known as **palaeontology**.

One of the major questions which has faced biologists has been to explain what has been responsible for evolutionary change. The idea that plants and animals have evolved is not new and is found in the writings of some of the philosophers in Ancient Greece. However, it was not until the 18th and 19th centuries that interest in evolution began to flourish and possible explanations for the causes of evolution were developed.

> **QUESTION**
>
> What features do trilobites show which enable palaeontologists to classify them as arthropods?

Natural selection

In 1859, **Charles Darwin** published a book which described the evidence for evolution which he had collected over 25 years. He called the book *The Origin of Species by Means of Natural Selection or the Preservation of Favoured Races in the Struggle for Life*. The ideas contained in this book are often referred to as **Darwin's theory of evolution by natural selection**, although another biologist, **Alfred Wallace**, working in south-east Asia, had independently developed the same ideas.

The theory of evolution by natural selection can be summarised by three key points.

- The individual characteristics of an organism, such as its height, colour, or speed of movement, are vitally important for its ability to survive and to breed.
- Individuals within a given species of organism vary in many characteristics. Individuals with certain advantageous characteristics, such as ability to avoid a predator, are likely to live longer and to produce more offspring.
- Only a small proportion of the offspring will survive. If those characteristics which help the organism to survive are inherited by their offspring, then

individuals with those characteristics will gradually become more common, generation after generation; in other words they will be **selected for**. The numbers of individuals less well adapted to the environment will correspondingly decrease, that is, they will be **selected against**. This is the basis of evolutionary change.

In the following sections we will consider some aspects of natural selection and look at the ways in which changes in the environment are likely to influence evolution.

Selection pressures and changes in allele frequency in populations

In Chapter 4, a **population** is defined as a group of individuals of one species found in the same habitat. Members of the same species are capable of interbreeding, so that it is possible for the exchange of genes to occur within a population. The term **gene pool** is used to refer to all the alleles of all the genes that occur within a population. The relative number of alleles of a particular gene within a population is referred to as the **allele frequency**, which is usually expressed as a percentage.

To take a simple example, consider a gene which exists as two alleles, **A** and **a**. Within a population, the genotypes **AA**, **Aa** and **aa** would be expected to occur (monohybrid inheritance, see page 80). We would expect the proportions of these genotypes to be in the ratio 1 : 2 : 1, so that within the population, we would expect 25% **AA**, 50% **Aa** and 25% **aa**.

This assumes, however, that the population is randomly breeding, and that there are no factors operating, such as mutation or natural selection, which may change these expected frequencies. Suppose, for example, that individuals with the genotype **aa** are selected **against**; this means that they are less likely to survive and pass on their **a** alleles to their offspring. Over a period of time, therefore, the frequency of the **a** allele within the population would be expected to decrease.

There are situations where both **AA** and **aa** individuals are selected against, so that the heterozygotes, **Aa**, have a selective advantage. An example of this is the inheritance of sickle-cell anaemia (see page 92) where people who are heterozygous for this condition (referred to as the sickle-cell trait) have a resistance to malaria and are therefore more likely to survive in areas where malaria is prevalent.

In Chapter 4, we described how populations, if unchecked, tend to increase in size. Some organisms are capable of producing enormous numbers of offspring. For example the weed, fat hen (*Chenopodium album*), may produce 400 000 seeds in one year. Of these, only very few will germinate and survive to reproduce. Some species of fish, such as tuna, produce up to 3 million eggs each time they spawn, but only one or two of these will survive to adulthood and the size of the population stays approximately constant. This is because growth of populations is normally kept in check either by increases in mortality (deaths) or by decreases in natality (births), as the population increases. The number of births or deaths

Figure 8.2 The forms of Biston betularia: *the* typica *form (top) seen on the bark of a tree from a rural area; (upper middle) the* carbonaria *form seen on lichen from a rural area; (lower middle) the* typica *form seen on the bark of a tree from an industrialised area; (bottom) the* carbonaria *form seen on the bark of a tree from an industrialised area*

is often expressed in terms of the number per adult member of the population. These figures are referred to as the birth rate, or **natality rate**, and the death rate, or **mortality rate**. As an example, a natality rate of 2 per cent would represent two births in a population of 100.

Availability of food is an important factor which can influence the natality rate. When food is plentiful, a population may increase dramatically in numbers as a result of an increase in the birth rate. This can lead to population explosions, as natality greatly exceeds mortality. Rapid increases in population size are frequently followed by an equally dramatic decrease as either the mortality rate increases or organisms move away from an area.

Increased numbers of prey organisms provide increased food for predators, and these can therefore be important agents which increase the mortality of prey. However, predators, such as birds, can be selective and this has a significant effect on the course of evolutionary changes in populations. For example, if some of the prey organisms are more conspicuous than others, they are more likely to be spotted by a predator and eaten. Less conspicuous organisms will therefore have a better chance of surviving to breed. This selective predation is an example of natural selection in action.

The peppered moth

The peppered moth (*Biston betularia*), has been the subject of very detailed scientific investigation. Studies on the peppered moth were started in the 1950s by H.B.D. Kettlewell and have been ongoing since that time. There are several different coloured forms of the peppered moth. The typical form, called *typica*, is pale and speckled. Another form, *carbonaria*, is a sooty-black colour. The moths are active at night, but settle on the bark of trees during the day. In rural parts of Britain where the air is cleaner, the bark of trees is lighter in colour than in the sooty areas of industrialised towns and is often covered by lichens and small plants such as mosses. These small plants are pale green or almost white in colour. More soot is deposited on trees in industrialised areas and fewer lichens and other small plants grow on these trees. Lichens and mosses are sensitive to sulphur dioxide in the air, which prevents their growth.

The pale (*typica*) form of the peppered moth is difficult to see against the bark of a tree in a rural area, but the black (*carbonaria*) form is very conspicuous. However, when viewed against the bark of a soot-covered tree, the reverse is true. Here the *typica* form stands out but the *carbonaria* form is almost invisible. These are illustrated in Figure 8.2.

Moths of both colours are eaten by birds such as thrushes, robins and nuthatches. It might be supposed that the more conspicuous forms are more likely to be predated by these birds and will therefore be selected against. Kettlewell used several different methods to investigate whether the colour of the moths made any difference to their ability to survive in different localities. He used each method in two areas: one a soot-covered woodland near Birmingham and the other an unpolluted woodland in Dorset. Two of these methods and the results are summarised below.

• ***Method 1:*** Equal numbers of both the *typica* and *carbonaria* moths were released, which settled on trunks and branches in the woodlands. The moths were watched and the numbers of each form that were eaten by birds were recorded. If, during one day's observation, all the moths of a particular form were eaten, more of that form were released to ensure that both forms were always available for the birds.

The total numbers of the different forms that were eaten by birds during 2 days' observation in each woodland are shown in Table 8.1.

Table 8.1 *Total numbers of peppered moths eaten by birds in two woodlands*

Colour form	Woodland near Birmingham	Woodland in Dorset
typica	43	26
carbonaria	15	164

- **Method 2:** Moths were captured in each woodland and the numbers of *typica* and *carbonaria* forms were counted. Two different trapping methods were used: a **light trap**, in which moths are attracted to a fluorescent lamp, and an **assembler trap**. In an assembler trap, female moths are placed in a gauze container and pheromones released by the females attract males, which can then be caught in a net. Using two different trapping methods in each woodland ensured that the samples caught were representative, as it is possible that one colour form might be more attracted to a light than the other form. The proportions of each form captured by each method were very similar.

The total numbers of moths caught by both trapping methods in the two woodlands are shown in Table 8.2.

Table 8.2 *Total numbers of peppered moths caught in two woodlands*

Colour form	Woodland near Birmingham	Woodland in Dorset
typica	55	359
carbonaria	422	34

The results of method 1 show that the more conspicuous colour form is eaten by birds in larger numbers in both areas. The results of method 2 indicate that the inconspicuous form is much more numerous than the conspicuous form in each woodland. Clearly the inconspicuous form is less likely to be eaten by birds than the conspicuous form.

The main conclusion to be drawn from these experiments is that peppered moths whose colour contrasts with their background are less likely to survive than peppered moths whose colour is similar to their background.

The story of the peppered moth illustrates an important principle in evolutionary biology: that **the phenotype of an organism affects its chances of survival**. As a general rule, the ability of an organism to survive will influence its ability to produce offspring; in other words, the longer an

ENVIRONMENTAL CHANGE AND EVOLUTION

Figure 8.3 Variation in the brown-lipped snail, Cepaea nemoralis

organism lives, the more likely it is to reproduce. The term **fitness** means the ability of an organism to survive and produce offspring which themselves can survive and produce offspring. In the case of the peppered moth, different phenotypes in one locality will differ in their fitness.

The brown-lipped snail

There have been many other studies of predators acting as selective agents, for example the influence of the song thrush on populations of snails. The brown-lipped snail (*Cepaea nemoralis*) exists in a number of different-coloured forms (Figure 8.3). The shell can be almost white, yellow or pink and there can be a variable pattern of brown bands around the shell. These forms, like the different forms of the peppered moth, are determined genetically. The main predator on *Cepaea* is the song thrush. Studies have shown, for example, that in beech woods, where the ground is covered by dark brown leaf litter, the pink-brown forms of *Cepaea* are better adapted to survive predation by thrushes than the more conspicuous yellow forms. By comparison, amongst grasses adjacent to sand dunes, the yellow form of this snail predominates.

Thrushes break the shells of snails by holding the snail in their beak and cracking it against a stone, referred to as an anvil. If you have access to such an anvil, in a garden, park or woodland for example, you could collect the shells and compare the numbers of different coloured forms which have been selected by thrushes.

Predation by thrushes is not the only selective agent acting on populations of *Cepaea*. In some exposed places for example, the light-coloured forms predominate although they are conspicuous. These forms, however, reflect the Sun's rays during the day and radiate less heat during cold nights, so climate can be a selective agent for shell colour.

Natural selection can therefore act in different ways on the same species in different parts of the organism's distribution. This is one way in which a species can diverge and, ultimately, develop into separate species in different places.

Stabilising, directional and disruptive selection

We have already stated that individuals within a particular species show variation in a number of inherited characteristics. This variation is important because natural selection acts against some individuals, leaving others to survive and reproduce. Variation in a characteristic, such as height or mass, shows **continuous variation**. When plotted as a histogram, continuous variables produce a bell-shaped curve, often showing a **normal distribution**. As an example, Figure 8.4 shows the distribution of lengths of a sample of 86 leaves. Notice that the lengths are expressed as ranges (for example, 120 to 130 mm) and the numbers of leaves (or frequency) in each range are plotted on the y-axis.

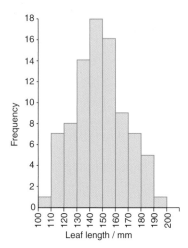

Figure 8.4 Histogram showing the frequency of leaves of different lengths in a sample

There are three ways in which natural selection can act on a population showing continuous variation, known as **stabilising**, **directional** and **disruptive selection**. These are illustrated in Figure 8.5.

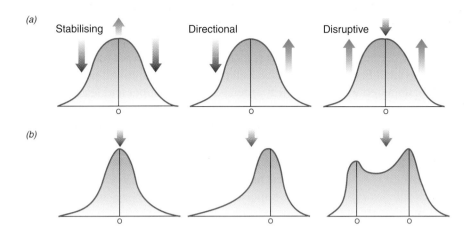

Figure 8.5 Three types of natural selection: (a) selection beginning; (b) distribution after selection. O = optimum value. Upward-pointing arrows indicate a selective advantage; downward-pointing arrows indicate a selective disadvantage.

Stabilising selection

This type of selection favours the mean of the distribution. One example of this is seen in the study by M.N. Karn and L.S. Penrose on the relationship between human birth weight and mortality. Karn and Penrose collected the birth weights of 13 730 babies born in a London hospital over a period of 12 years together with data on the survival of the babies. Figure 8.6 shows the relationship between mortality and birth weight, mortality being determined by the percentage of babies failing to survive for 4 weeks. This study shows that the optimum birth weight (that is, at which mortality is lowest) is very close to the mean value. On either side of this value, the expectation of survival with either increasing or decreasing birth weight decreases rapidly, reaching a minimum at the two extremes of the distribution.

Directional selection

This type of selection favours one extreme of the range of characteristics, as the other extreme is selected against. Examples of directional selection include

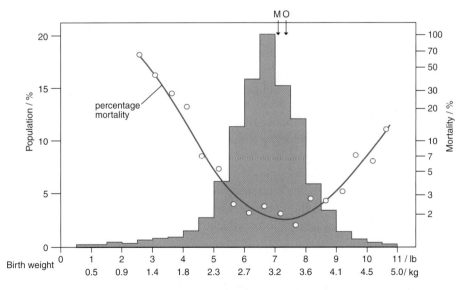

Figure 8.6 Relationship between infant mortality and birth weight. M = mean birth weight; O = optimum birth weight.

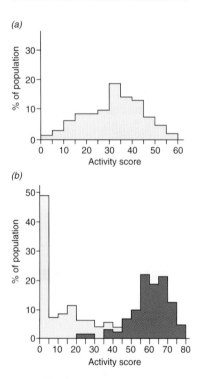

Figure 8.7 Spontaneous activity in Drosophila: *(top) frequency distribution in original population before selection; (bottom) frequency distribution after selection for 14 generations*

selection against light-coloured peppered moths in an industrialised area, and selection against grasses which are intolerant of heavy metals in soil contaminated with lead, tin or zinc.

Varieties of certain species of grasses, including *Agrostis tenuis* (common bent-grass) and *Festuca ovina* (sheep's fescue) have been found growing on soil contaminated with ions of heavy metals, such as copper and lead. These soils may occur naturally, or be the result of mining activities and metal ore extraction. These metal ions are normally toxic to plants, but those that are tolerant are able to grow, in other words, they have been **selected for**. Varieties that are not tolerant of the heavy metals are unable to establish on contaminated soil, so are **selected against**. It is likely that, over time, the varieties will diverge into separate species: those that are heavy metal tolerant, and those that are not. This is an example of sympatric speciation (page 113).

Disruptive selection

Disruptive selection favours both extremes of a distribution, with selection occurring against the mean. This eventually results in a bimodal distribution, as shown in Figure 8.5. Disruptive selection has been demonstrated in both natural populations and laboratory experiments. For example, populations of the fruit fly (*Drosophila*) show variation in the amount of spontaneous activity when each fly is placed in standardised conditions. Figure 8.7 shows the result of an experiment in which males and females from the low activity end of the activity range were mated together, and males and females from the high activity end of the distribution were mated. After 14 generations, two distinct populations emerged.

In natural populations, disruptive selection may select against intermediate forms. For example, some individuals of the European swallowtail butterfly (*Papilio machaon*) pupate on leaves or stems which are brown; others pupate on green leaves or stems. Two distinct colour forms of the pupae are found, namely brown and green, but very few intermediates.

Isolation and speciation

Populations of organisms are genetically very variable, and one population may diverge to form two separate species in the following way. Suppose that there is a population of moths living in all parts of a tropical forest. The conditions in the forest have been very similar for millions of years and the moths are well adapted to these conditions. Now suppose that the climate changes and generally becomes much drier. Much of the forest will die out and be replaced with scrub or desert. Some of the forest, however, survives as two separate areas on mountain sides, which remain sufficiently wet to maintain the forest. What was previously a widespread, single population of moths is now separated into two populations, confined to the forest regions on the mountains. The moths which previously lived in the areas of forest which have disappeared have died out.

It is likely that the climatic conditions on the two mountains will be different. One might be colder and wetter than the other, as illustrated by Figure 8.8. Over many generations, natural selection would favour genetically new kinds of moths which would be better adapted to these new conditions. Where

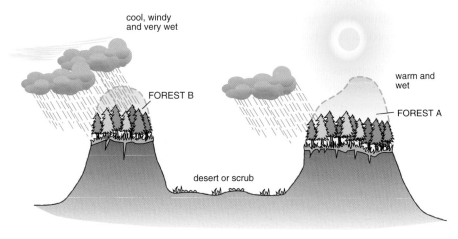

Figure 8.8 Different climates in two forest areas which were once part of a single area of forest

previously there was just one kind of moth, there would now be two, which are genetically distinct.

Suppose that the climate now changes again and the low-lying desert area becomes wetter and tree-covered. The moths in the two populations are now able move into the newly grown forest and meet up. If the two populations have not diverged too much, they might mate with each other, produce fertile offspring and merge into a single population again. It is possible, however, that the two populations will have diverged so much that they are unable to interbreed; in other words they have become **reproductively isolated** from each other and therefore can be said to be different species. Once two populations become reproductively isolated they evolve along their own separate paths.

Geographical isolation and allopatric speciation

In the example we have described, the cause of speciation of the two populations of forest moths was **geographical isolation** arising from a change in the climate. Speciation as a result of geographical isolation is also referred to as **allopatric speciation** (different places). Divergence and speciation in populations may also arise as a result of **behavioural isolation**. For example, small differences in the feather patterns or songs of some birds may change their attractiveness as mates to members of their original population. Similarly, plants living in the same area may be reproductively isolated if they flower at different times of the year. Small differences can therefore isolate a group from the rest within one population, and natural selection will increase the divergence until they constitute a new species.

Sympatric speciation

Sympatric speciation occurs within an existing population (in the same area) and arises if, for example, mutant individuals preferentially mate with each other, or if sterile hybrids are formed. In the latter case, the parental types will diverge. As an example, polyploidy in plants may immediately isolate a polyploid plant from others. Many plants are capable of asexual reproduction, so the

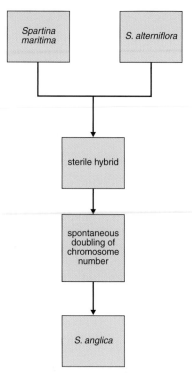

Figure 8.9 Evolution of a new species of cord grass by interspecific hybridisation

numbers of polyploid plants may increase until a large, interbreeding population is formed. Cross-breeding between, for example, diploid and tetraploid plants would result in the formation of sterile triploid hybrids, so isolation is produced between diploids and tetraploids.

This can be illustrated by reference to the evolution of a new species of cord grass. Cord grasses are a group of plants which colonise mudflats in estuaries. The original native British species is the small cord grass, *Spartina maritima*. In 1878, there was a report of a new, interspecific hybrid between *S. maritima* and an accidentally introduced American species, *S. alterniflora*. Spontaneous doubling of the chromosomes of the hybrid resulted in a new species, *S. anglica*, which is self-fertile, but sterile with both of its original parental species. *S. anglica* is well adapted to living in estuarine conditions, having a vigorous growth and a good ability to withstand fluctuating levels of salinity. The evolution of *S. anglica* is shown in Figure 8.9.

EXTENSION MATERIAL

Prezygotic and postzygotic reproductive isolation

Genetic isolation is an essential requirement for speciation. Isolation of one population of individuals may result in the accumulation of different allele frequencies and they may eventually behave as a separate species. The mechanisms for maintaining genetic isolation between populations of one species are known as reproductive isolating mechanisms. Prezygotic (or premating) isolation means that zygotes are not formed because gametes do not meet. However, if zygotes are formed, they may fail to develop as a result of postzygotic (or postmating) isolation.

Table 8.3 lists some of the ways in which reproductive isolation may be brought about.

Table 8.3 *Reproductive isolating mechanisms*

Time when isolation is effective	Nature of isolation	Notes
prezygotic	geographical	populations inhabit different areas
	ecological	populations use different habitats within one environment
	temporal	populations inhabit same area but are active or reproduce at different times
	behavioural	courtship displays prevent mating between individuals of different species
	mechanical	reproductive parts may not fit each other
postzygotic	hybrid inviability	hybrids are produced but they fail to live to maturity
	hybrid sterility	hybrids fail to produce functional gametes
	hybrid breakdown	F_1 hybrids are fertile but F_2 generation fails to develop or is infertile

Gene technology

Gene technology involves the manipulation of genetic material so that, genes from another organism can be inserted into cells. Such altered genetic material is referred to as **recombinant DNA**. Many of the techniques used in gene technology were originally developed using bacteria, particularly *Escherichia coli* (*E. coli*), and enable the insertion of mammalian genes into bacterial cells. This may result in the large-scale production of proteins in bacterial cultures. The first of these to be licensed for human use was insulin, in 1982.

The basis of recombinant DNA technology was established in the 1970s, with the discovery of several enzymes that enable DNA molecules to be cut, copied and joined. These enzymes include restriction endonucleases (type II), DNA ligase and reverse transcriptase.

The manipulation of DNA

Restriction endonucleases are extracted from microorganisms. They get their name because they 'restrict' invasion of the host cell by foreign DNA molecules, such as viral DNA, by cutting it up. These enzymes cut at sites within the foreign DNA, rather than removing bases from the ends, so they are called endonucleases, as opposed to exonucleases. Hundreds of different restriction endonucleases have now been isolated and type II restriction enzymes are particularly useful because they cut DNA molecules at specific nucleotide sequences, known as **recognition sites**. Two such enzymes are known as Eco RI and Hind III, which cut DNA at the specific recognition sites (Figure 9.1).

Notice that each enzyme makes a staggered cut, leaving two complementary ends. These complementary ends have a natural affinity for each other, and are referred to as 'sticky ends'.

Once a desired gene has been isolated, it is introduced into a host cell using a vector, such as a **plasmid**. Plasmids occur in bacterial cells and consist of small, circular loops of DNA, quite separate from the bacterial 'chromosome'. They are widely used as vectors in gene technology because of their relatively small size and their ability to replicate within bacterial cells. Once in the host cell, the plasmid can produce many copies of the original gene. Plasmids are used as **vectors** to introduce a gene of interest into a host cell. Once in the host cell, the plasmid will replicate so that many copies of the original gene will be produced in each cell.

Plasmids can be cut open using a restriction endonuclease. The desired gene is cut out from another organism's DNA, using the same restriction endonuclease. The 'sticky ends' on this gene and the 'opened' plasmid will be complementary. This means that the gene and the plasmid will join easily by hydrogen bonding between complementary bases. An enzyme called **DNA ligase** will join the gene and the plasmid permanently by forming covalent phosphodiester bonds (Figure 9.2).

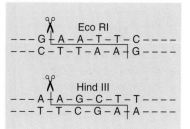

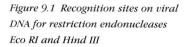

Figure 9.1 Recognition sites on viral DNA for restriction endonucleases Eco RI and Hind III

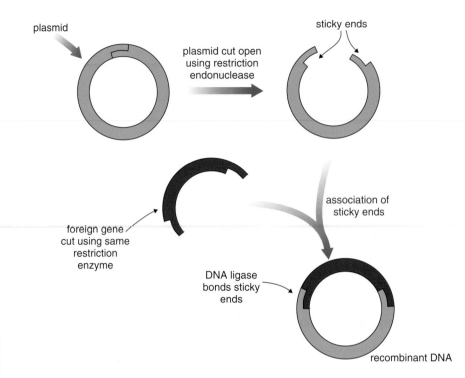

Figure 9.2 Formation of recombinant DNA from a bacterial plasmid acting as a vector for a 'foreign' gene, such as the gene for human insulin

The recombinant DNA is then placed into bacterial cells. This can be achieved by treating bacterial cells with calcium ions, which makes their membranes more permeable, so they will take up DNA if it is added to their growth medium. Bacterial cells that have taken up the recombinant DNA are referred to as **transformed cells**. Treated cells are spread out on a suitable medium in a Petri dish so that each cell will grow into a separate colony, or clone. Some of these cells will contain the vector only, if the plasmid has failed to take up the foreign DNA, and others will contain the gene of interest.

Usually a marker, such as a gene for antibiotic resistance, is inserted into the plasmids in addition to the desired gene. When the transformed cells are cultured, it is possible for them to grow on a medium containing the antibiotic, whereas cells which have not taken up the plasmids will not grow. In this way, the transformed cells can be selected.

Another method of finding transformed cells involves the use of a radioactive DNA probe containing part of the sequence of the gene required. This probe is used to find cells that have DNA which can hybridise with it. These cells are then subcultured to provide a pure culture of transformed cells, which can be grown on a large scale (up to 10 000 dm³) in industrial fermenters. Bacteria multiply quickly and can be grown in relatively cheap media. The transformed cells will synthesise the protein product, which is then extracted from the cells.

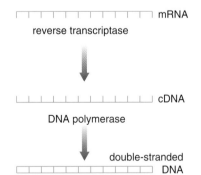

Figure 9.3 Use of the enzymes reverse transcriptase and DNA polymerase in the production of double-stranded DNA from a mRNA template

Reverse transcriptase is an enzyme that was first isolated from viruses in 1970. This enzyme will form DNA from an RNA template. This allows complementary, or copy, DNA (cDNA) to be formed from an extract of mRNA. The outline of this process is shown in Figure 9.3.

This technology is used in the production of human insulin. Messenger RNA, with the codons for insulin synthesis, is isolated from pancreatic cells. Reverse transcriptase allows a complementary strand of DNA (cDNA) to be made from the mRNA. The cDNA is made into double-stranded DNA sequences, which are inserted into a suitable plasmid vector, which can be used to transform bacterial cells.

The ability to synthesise a specific protein in large quantities has considerable medical, industrial and agricultural potential. A number of such proteins produced by recombinant DNA technology and now in routine use include human growth hormone, erythropoietin (a growth factor secreted by the kidneys that stimulates the production of red blood cells and is used in the treatment of certain types of anaemia), blood clotting factors and bovine growth hormone, or bovine somatotrophin (BST).

Introducing new genes into crop plants

The soil bacterium *Agrobacterium tumefaciens* causes crown gall disease in dicotyledonous plants. This bacterium gains entry to the plants through a wound and stimulates the production of a tumour, referred to as a gall, in the stem. The growth of the tumour tissue is due to the presence in the bacterium of a plasmid, the T_i plasmid. A piece of this plasmid can become incorporated into the DNA of the host plant cells, where it replicates and causes the host plant to release hormones, which stimulate the production of the cells forming the tumour. This plasmid can be isolated, its tumour-inducing gene removed and the desirable genes introduced into it. The plasmid is then used as a vector to carry the desirable genes into the plant issue. The resulting callus (undifferentiated tissue) produced by the plant cells can be induced to **transgenic plants** containing the desirable genes (Figure 9.4).

Producing chymosin from genetically modified yeast

In cheese making, **chymosin** (rennin) is the main enzyme involved in the coagulation of casein, the protein in milk. Traditionally, the source of chymosin was **rennet**, an extract from the abomasum (stomach) of young calves, or sometimes from kids or lambs. In the 16th century, rennet was prepared by cutting strips of the stomach of young calves and steeping these in warm milk or brine to extract the rennet. By the late 19th century the first industrial preparation of calf rennet was established by a Danish chemist. Calves destined for consumption as veal were used, so they were not sacrificed specifically for the extraction of the enzyme. More recently, in the 1960s, because of changing eating patterns, there was concern that there would be a world-wide shortage of rennet for commercial cheese production. This led to pressure to find alternative sources of rennet and to develop substitutes to keep up with the demand.

Bovine rennet from adult cattle can be used as an alternative to calf chymosin, but the bovine extract contains a higher proportion of pepsin and gives a lower yield of cheese. Certain fungi produce proteases which can clot milk proteins. **Fungal enzymes** are now used in more than one-third of cheese produced world-wide. Three fungi used for production of the enzymes are *Mucor miehei*, *M. pusillus* and *Endothia parasitica*. Compared with calf chymosin, the fungal enzymes are more stable, but this can be a disadvantage in cheeses which have a long maturing stage (for example, Cheddar cheese) because degradation of the

Figure 9.4 Formation of transgenic plants

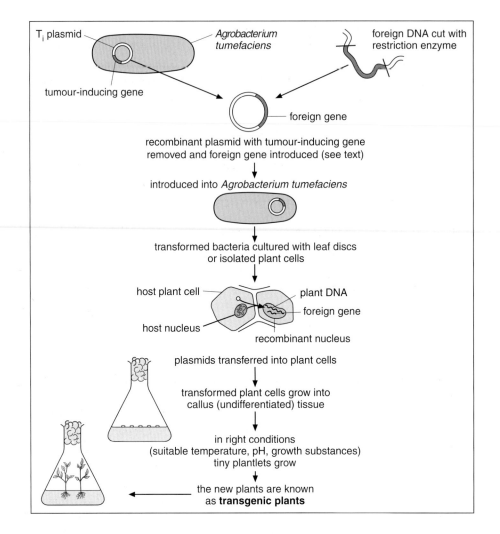

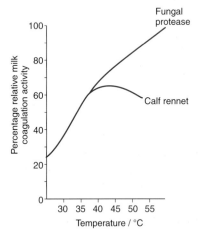

Figure 9.5 Differences in behaviour of calf rennet and fungal protease – influence of temperature

milk proteins continues. To counteract this, these enzymes can be destabilised, using oxidising agents, so that they behave in a way similar to the more successful calf chymosin. Fungal enzymes are used widely in production of cheese for vegetarians (Figure 9.5).

DNA technology has provided further substitutes for calf rennet. The first microorganisms capable of making chymosin were produced in 1981, using *Escherichia coli*. Now chymosin is produced from genetically modified yeasts, including *Kluyveromyces lactis* and *Saccharomyces cerevisiae*. Precisely the same DNA code as in the calf is incorporated into the microorganism, so the enzyme produced is identical to that from calves (Figure 9.6). Expert tasters can detect no differences between the cheeses produced using chymosin from genetically modified organisms and that from extracted calf rennet. The enzymes actually have fewer impurities and their behaviour is more predictable. At first there was resistance to accepting cheese made with the involvement of genetically modified organisms (GMOs). Before being released for general consumption, there was rigorous testing of the products. The enzymes used for cheese produced in this way have been approved by the relevant regulatory bodies and by the Vegetarian Society. Such cheese is on sale in several countries, including the UK.

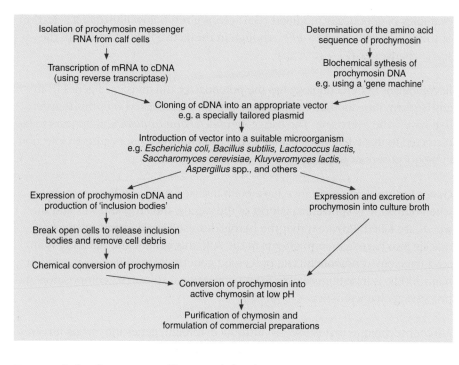

Figure 9.6 Stages in the production of calf chymosin by genetically modified microorganisms. Prochymosin is an inactive precursor of chymosin.

Potential of genetically modified organisms

For centuries, farmers have modified their plant crops and their domesticated animals. They have done this by the slow processes of breeding and selection of desired characteristics in the progeny and have thereby manipulated the gene pool and achieved considerable improvement in the varieties used. This selective breeding is still the most frequent method of developing new varieties. However, modern biotechnology carries conventional methods of breeding and artificial selection a stage further, and the techniques of genetic modification are now being applied to the breeding of crop plants. Figure 9.7 illustrates ways in which conventional and GM techniques have been applied to the improvement of wheat and bread.

With its 'cut and paste' techniques, DNA technology (gene technology) allows much more precise control of the specific genes that are incorporated into the genome and success is likely to be achieved in a much shorter time. Some of the aims in current research programmes are outlined in Table 9.1. The few

Table 9.1 *Current research with genetically modified crop plants and possible benefits*

Feature in genetically modified organism	Possible benefits
improve efficiency of uptake of mineral salts	reduced fertiliser input
improve ability to withstand drought or high salt	crop can be cultivated on land where soil or climate is unsuitable
improve resistance to herbicides	crop better able to survive application of herbicides to control weeds
improve resistance to disease	reduced pesticide input, crop losses reduced
improve frost resistance	growing and harvest season extended
control ripening of fruits	postharvest losses reduced

DEFINITION

Genetic modification in relation to an organism means the altering of the genetic material in that organism by a way that does not occur naturally by mating or natural recombination or both.

examples of genetically modified (GM) crops given below illustrate a selection of the achievements up to 1997, though in this active field of research, the scene changes fast.

Recombinant DNA technology has the potential for the transfer of genes from any life form into another, but there are problems associated with this transfer that have to be overcome. The use of *Agrobacterium tumefaciens* appears to be restricted to dicotyledonous plants, as the bacterium does not cause an infection in monocotyledonous plants such as grasses and cereal crops.

Genes for herbicide resistance have been inserted into soya beans, so that when herbicide is sprayed to get rid of the weeds, the crop survives and the weeds are killed. Provided that the herbicides used break down relatively quickly into harmless components in the soil, this genetic modification is an advantage environmentally. The processed genetically modified soya, used in many foods, is indistinguishable from conventional soya in its composition and processing characteristics.

Another technique that is used to introduce 'foreign' genes into crops involves the use of plant cells that have had their cellulose cell walls removed. Treatment with cellulase enzymes removes the cell wall, leaving protoplasts which can be cultured in suitable media. The protoplasts will regenerate new cellulose cell walls in time, but it is possible to introduce new genes, through the plasma membrane, before this occurs.

The bacterium, *Bacillus thuringiensis*, produces a protein which is toxic to a variety of insect larvae, but harmless to other animals and humans. The gene for the production of this protein is located in the plasmids of the bacterium. Copies of these bacterial plasmids can be induced to enter the protoplasts of cells from crop plants. The cells are then cultured to form a colony and insect-resistant plants can be produced. This technique has been used to produce genetically-modified maize plants which can synthesise the toxic protein and possess a natural defence against insect pests such as the European corn borer. It has been estimated that this pest destroys around 4 per cent of maize crops world-wide, by boring through the stems and ears of the plants, causing them to fall over.

Rice and resistance to disease

Rice is a very important crop on a global scale. It suffers from the rice stripe virus (RSV). Transgenic rice, produced by introducing the gene for the virus protein coat into the rice plant genome, shows noticeably increased resistance to the rice stripe virus.

Genetically modified tomatoes

The use of gene technology to improve crop quality is demonstrated by the development of genetically modified tomatoes. Tomatoes soften as they ripen, mainly because of the activity of the enzyme polygalacturonase (PG). PG acts on pectic substances in the middle lamella within the cell wall structure, hydrolysing long polymers and converting them to shorter, more soluble fragments. The cells lose their cohesion, so begin to move in relation to each other. Turgor of the cells diminishes as the cell walls weaken, resulting in loss

QUESTIONS

'Cut and paste' techniques with DNA

Different techniques are used in the manipulation of DNA. Find out what you can about the following techniques:
- gene synthesis and the polymerase chain reaction
- using plasmids as vectors
- gene ballistics
- electroporation
- microinjection of DNA (into newly fertilised eggs)
- cell fusion.

What organisms is each technique used for – plant, animal or microbe?

(a)

(b)

Figure 9.7 Improvement in wheat plants: (a) conventional plant breeding has produced varieties of wheat with different stem length – the shorter, stronger stems help prevent the wheat plants from being blown over in wet, windy conditions; (b) improved bread flour from genetically modified wheat plants. Genes for high molecular weight (HMW) polypeptides have been transferred experimentally into certain wheat cultivars and the bread made from this flour (left) has a lighter, spongier texture than that produced from flour low in HMW polypeptides (right).

of firmness. Synthesis of PG coincides with a rise in ethene concentration and ethene is involved in promoting the ripening process. Inhibition of PG activity slows the softening but still allows development of the desirable flavours and colour associated with ripening.

Genetic modification, using two different methods, has led to a precise way of reducing the expression of the PG gene so that the tomato remains firm. The gene that codes for PG has been identified and sequenced. Using *Agrobacterium* as a vector, an 'antisense' PG gene has been inserted into the tomato plant (Figure 9.8). An antisense gene is effectively a reversed form of the gene and is inherited in a normal Mendelian way. Transgenic tomatoes produced by antisense technology have been approved for sale in the USA, under the name of '*Flavr-Savr*'. However, production of Flavr-Savr tomatoes was short-lived, due to problems with disease resistance and harvesting. Research in the UK has produced genetically modified tomatoes with a shortened PG gene. In the UK, approval for the sale of paste (but not the fresh fruit) made from these genetically modified tomatoes was given in 1995.

As well as improved flavour, these tomatoes show improvements in consistency and viscosity, giving a thicker tomato paste, without the need for addition of thickeners. There is much less wastage in the field at harvest and in the processing. It is of interest that in one supermarket chain in the UK, over three quarters of a million cans of GM tomato paste were sold from its introduction in February 1996 up to November 1997. The cans were clearly labelled to indicate the paste had been made with GM tomatoes and the conventional equivalent was always available alongside. In some stores, sales of the GM paste exceeded that of the conventional paste.

GENE TECHNOLOGY

Figure 9.8 Two routes to genetically modified tomatoes

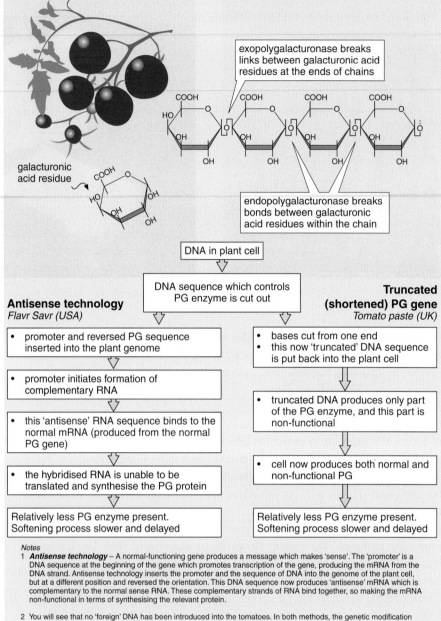

exopolygalacturonase breaks links between galacturonic acid residues at the ends of chains

galacturonic acid residue

endopolygalacturonase breaks bonds between galacturonic acid residues within the chain

DNA in plant cell

DNA sequence which controls PG enzyme is cut out

Antisense technology
Flavr Savr (USA)

Truncated (shortened) PG gene
Tomato paste (UK)

- promoter and reversed PG sequence inserted into the plant genome

- bases cut from one end
- this now 'truncated' DNA sequence is put back into the plant cell

- promoter initiates formation of complementary RNA

- truncated DNA produces only part of the PG enzyme, and this part is non-functional

- this 'antisense' RNA sequence binds to the normal mRNA (produced from the normal PG gene)

- cell now produces both normal and non-functional PG

- the hybridised RNA is unable to be translated and synthesise the PG protein

Relatively less PG enzyme present. Softening process slower and delayed

Relatively less PG enzyme present. Softening process slower and delayed

Notes
1 ***Antisense technology*** – A normal-functioning gene produces a message which makes 'sense'. The 'promoter' is a DNA sequence at the beginning of the gene which promotes transcription of the gene, producing the mRNA from the DNA strand. Antisense technology inserts the promoter and the sequence of DNA into the genome of the plant cell, but at a different position and reversed the orientation. This DNA sequence now produces 'antisense' mRNA which is complementary to the normal sense RNA. These complementary strands of RNA bind together, so making the mRNA non-functional in terms of synthesising the relevant protein.

2 You will see that no 'foreign' DNA has been introduced into the tomatoes. In both methods, the genetic modification has been achieved by manipulation of the tomato plant's own DNA.

Implications for the development of genetically modified plants and products

The impact of modern biotechnology raises issues beyond just the science and technology. A common initial reaction to DNA technology is that it is 'unnatural', 'interfering with nature'. There is the feeling that people have gained too much control over the fate and future of living organisms, that they are 'playing God'. For some religious groups there is belief in the fixity of created species, and such people find it difficult to accept any sort of interference or manipulation at the genetic level. There are suggestions that biotechnology involves taking risks for commercial gain, and that the biotechnology companies benefit rather than some of the economically vulnerable communities in the world. These are ethical and moral concerns and arouse strong and different emotions in people. The debate is carried out by a chain of people involved in the decision making – those responsible for the original research, the field trials, the food companies and retailers with their marketing strategies through to the consumer or wider public, and the regulators who scrutinise the development at every stage.

There are fears that release of GMOs could be directly harmful to humans (or other organisms), or disturb the ecological balance and natural interaction between organisms – a balance that has evolved over a long period of time. Less extreme is the possibility of 'biological pollution' – that GMOs, once released, will spread and compete with or destroy existing wild populations. Reservations about DNA technology are stronger when it has been applied to food or to farm animals, but acceptance is greater in the field of medicine or where plants are involved.

Research with GMOs is done under controlled and confined conditions. In the development stages, genetically modified crops undergo extensive field trials before there is any question of releasing the GMO. The research tries to predict the possible influence of the GMO as well as assess any risk of harmful ecological or environmental change. Such trials are carried out in many locations, and each transgenic crop is considered on an individual basis. Similar procedures are carried out for transgenic animals and microorganisms.

In the UK, all foods, before being released for public sale, have to meet a range of requirements, as laid down in the Food Safety Act, 1990. There are committees and regulatory bodies which look closely at novel foods and processes, including those involving genetically modified organisms. The Advisory Committee on Novel Foods and Processes is one such committee which has looked at a range of GM-derived foods, including tomato paste, herbicide-tolerant soya beans and chymosin from microbes. Members of this committee include research scientists, representatives from the food industry, consumer and environmental groups and medical experts. In considering GM foods, it is generally agreed in the UK that, in terms of safety, the food should be compared with an analogous conventional food to establish whether it is 'substantially equivalent'. It is also agreed that companies offering GM foods for sale should be encouraged to provide information about the product or ingredient which is readily accessible to the consumer. Such information can be supplied by means of the product label or by information leaflets available at the time of purchase. Provided appropriate information is available and there is choice between GM and conventional food, it can then be an individual decision for each person to decide whether or not to consume GM food.

A biotechnology advisory working group (BAWG), comprising representatives from consumer organisations, agriculture, research, the major retailers and the food and related industries, has debated the issues at some length. The group has developed 'an extensive information framework, including Guidelines for communication and for labelling of genetically modified foods'.

> ## QUESTION
>
> For centuries people have eaten rice from plants infected with RSV (rice stripe virus). Do you think the same people would be happy to eat transgenic rice which contains the gene for the virus protein coat?

GENE TECHNOLOGY

QUESTIONS

- Do you think that herbicide resistant crop plants would encourage more spraying with herbicides or less use of herbicides?
- What benefits might there be if the ability to fix nitrogen is transferred to cereal crops?
- Think about the economics, possibility of growing crops on marginal land, and environmental consequences in relation to use of fertilisers.

In a statement made in 1997, BAWG supports a recommendation from the Food Advisory Committee, that food should:

> *'be labelled if it contains a copy gene originally derived from: a human; an animal, the eating of which is restricted by some religions such as pigs for Muslims and cows for Hindus; or if it is a plant or microbial food and is modified with genes originally derived from any animal'.*

The major issues to be considered are the concerns about the possible impact of genetically modified organisms on the environment and to what extent the technology may improve the quality of life. Public debate raises a number of concerns about the use of GM crops in particular, which can be summarised as follows:

- the possiblility of cross-pollination between GM and non-GM crops, leading to contamination of seeds and other crops
- the implications for organic farming
- the spread of herbicide resistance from crops to weeds
- the effectiveness of pest resistance in the long term
- the possibility of contamination of human food and the long term effects on human health

Some of these concerns can be dispelled by an understanding of the life cycles of the organisms involved, but it is also worth considering the possible benefits to agriculture in order to gain a balanced view. These include:

- less use of pesticides and herbicides so that fewer toxic chemicals are released into the environment
- improvement in the quality of food means less wastage
- improvement in productivity means there is more food
- more land can be made productive by producing GM crops that can grow in less fertile areas
- continuation of research into the improvement of crops, e.g. introduction of genes for nitrogen fixation into cereal crops.

The polymerase chain reaction (PCR)

Detective work may have to depend on only minute quantities of material for use as clues in an investigation. The mystery to be solved may be concerned with forensic analysis, DNA fingerprinting, comparative evolutionary studies, prenatal detection of genetic abnormalities or diagnosis of disease. The material might be a single hair, a drop of blood, just one cell, viruses in clinical samples, or a fragment of ancient DNA from mammoths preserved in ice, extinct plants or mummified human remains. The PCR provides an *in vitro* means of amplifying (copying many times) a specific section from a sample of DNA, thus increasing the amount of material which can then be used for further investigation. The PCR copies are available in just a few hours, whereas making copies through the technique of gene cloning in bacteria would take several days.

The target length of DNA to be copied is selected using **primers**. These are short lengths of about 20 to 30 nucleotides which are artificially synthesised. The primer nucleotide sequence precisely complements the nucleotide sequence which lies at each end of the target DNA. Two primers are used –

one is complementary to the target strand and the other is complementary to the opposite (complementary) strand of the DNA. The PCR reaction mixture contains the sample of DNA, an excess of appropriate primers, a **DNA polymerase enzyme**, the four **(deoxy)nucleotide phosphates** (dCTP, dATP, dGTP, dTTP – to build the new DNA) and **buffer**. The PCR cycle involves heating the reaction mixture to separate the strands of DNA, allowing the mixture to cool so that the primers anneal (join) to their respective strands, heating again to allow a fresh DNA strand to be synthesised extending from the primer position, then further heating to separate these newly formed strands of DNA. The exact temperatures used depend upon the base sequence and the length of the primers. The cycle is repeated many times, and each time the number of copies of the target length of DNA is doubled.

When first described in 1985, the PCR technique was carried out manually and was a rather lengthy process. The reaction mixtures had to be transferred manually through the series of water baths and fresh polymerase enzyme was added at each new cycle. The procedures were repeated 20 or 30 times for each cycle. Development of automated PCR machines has allowed much more widespread use of this technique for amplifying DNA, and its application in a variety of fields outside the research laboratory. Further developments of simpler, smaller-scale machines means it is probable that PCR machines will become available for use with A-level practical work in the near future, though the process can be carried out manually (but is tedious). The technique requires a polymerase enzyme which is active at the high temperatures

BACKGROUND

Summary of steps in the DNA-copying procedure (Figure 9.9)

- Heat the sample of DNA to 95 °C for about 20 seconds – the hydrogen bonds between the strands break, so the complementary strands of the DNA separate and the DNA is **denatured** into single strands.

- Cool to between 55 and 60 °C for 20 seconds – this allows the primers to anneal (join at the hydrogen bonds) to the complementary portion of the DNA. Using an excess of primers ensures that some bind with the DNA rather than the original strands joining again.

- Heat to 72 °C for 30 seconds – the DNA polymerase binds at the end of the primer, enabling free nucleotides to build a complementary strand along the exposed portion of the DNA thus **extending** the primers and restoring the double strand of DNA.

- Repeat the cycle 20 or more times – each time the target length of DNA which is copied becomes the template for the next cycle.

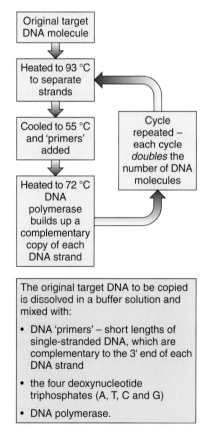

Figure 9.9 An outline of the polymerase chain reaction (PCR)

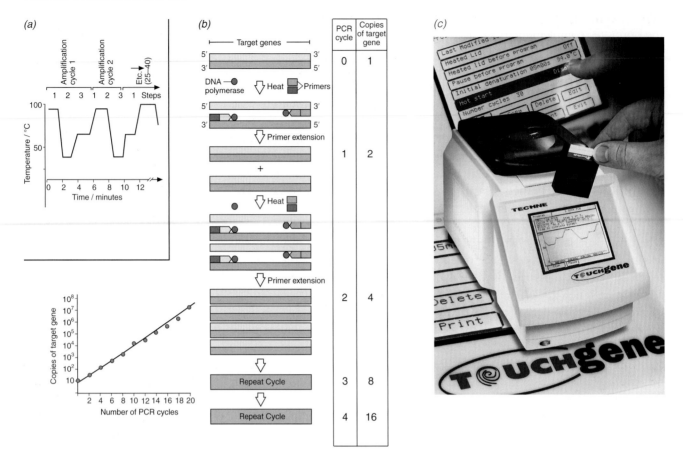

Figure 9.10 *PCR process: (a) temperature changes during each PCR cycle;*
(b) summary of the PCR process, showing how the DNA doubles at each cycle;
(c) a thermal cylinder or automated PCR machine for use with the PCR process.
The graphical display shows the stage of the cycle and the sample temperature.

involved in the cycle. A useful source of a heat-stable DNA polymerase has
been the bacterium *Thermus aquaticus*, which is stable above 95 °C. This is
known as *Taq* polymerase. As with any DNA work, it is important to avoid
contamination of the target sample with DNA from people handling the
material.

The sequence of events in the polymerase chain reaction is summarised in
Figure 9.10. You can see that, in four cycles, a single length of the target DNA
produces 16 copies. Work out how many copies would be produced in
20 cycles – see if it exceeds 1 million! You will begin to appreciate that the PCR
is a powerful tool to use in generating enough DNA to do further analysis.
Looking for a single gene has been described as 'searching for a needle in a
haystack'. Using the PCR has now been dubbed 'making a haystack out of
needles'. The PCR is also referred to as a means of carrying out 'molecular
cloning'.

Genetic fingerprinting as a diagnostic tool

A large percentage of the human DNA does not code for proteins and appears to have no function. Random mutations occur in this non-functional DNA and these accumulate, so that there are large numbers of differences between the DNA of unrelated individuals. Comparisons of the DNA of different individuals can be used in criminal investigations, to determine relationships and in the diagnosis of genetic disorders (Figure 9.11).

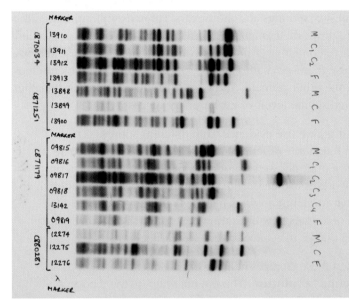

Figure 9.11 An example of a genetic fingerprint

When human DNA is treated with restriction endonucleases, it is split into fragments which can be separated into bands using gel electrophoresis. The fragments separate according to their size and charge: the larger the fragments, the more slowly they move. At this stage, individual bands do not show up on the gel.

Different methods are used to show up the bands, but typically the DNA is allowed to react with a radioactive DNA probe. Complementary probes bind by hydrogen bonding and an X-ray film is used to obtain an autoradiograph, known as a 'fingerprint' (Figure 9.12). This fingerprint will be unique to the individual from which the DNA was taken.

Genetic fingerprinting is widely used by forensic scientists to help in the solving of crimes. Single hair roots, small splashes of blood or minute quantities of cells can provide enough DNA for analysis, as it is possible to copy the DNA using PCR techniques.

Genetic fingerprints can also be useful in diagnosing genetic disorders such as Huntington's chorea, sickle-cell anaemia and β-thalassaemia. Each of these conditions is due to a single abnormal allele. By obtaining a few fetal cells, from amniocentesis or chorionic villus sampling, and cutting up the DNA with the appropriate restriction endonucleases, it is possible to see whether a child has inherited a normal or a mutant allele.

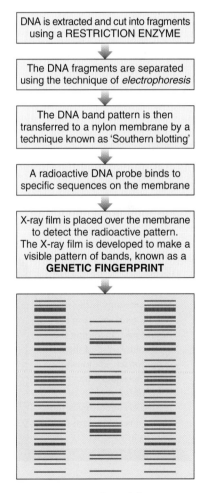

DNA is extracted and cut into fragments using a RESTRICTION ENZYME

The DNA fragments are separated using the technique of *electrophoresis*

The DNA band pattern is then transferred to a nylon membrane by a technique known as 'Southern blotting'

A radioactive DNA probe binds to specific sequences on the membrane

X-ray film is placed over the membrane to detect the radioactive pattern. The X-ray film is developed to make a visible pattern of bands, known as a **GENETIC FINGERPRINT**

Figure 9.12 An outline of the stages used to produce a genetic fingerprint

Human evolution

Introduction – origins and evolution of humans

This chapter attempts to unravel the biological events in the story of how humans evolved to become the dominating mammal on Earth. Humans are animals capable of using tools, who cultivate land to produce food, and have brains that create music, paintings and motor cars. Humans walk upright, talk, sing songs and write down their language. Humans build houses, wear clothes and exert considerable control over their environment. They care for their children, and organise themselves into social groups with elaborate hierarchies and interactions. While other primates show some of these attributes, no other mammal has achieved similar levels of evolutionary success.

To reconstruct the story, we use evidence from different sources. This evidence is often in fragments or indirect and while we sift and evaluate it systematically and scientifically, we cannot do experiments to confirm our findings. Research workers use modern techniques to analyse the material, even at the molecular level, but answers may be limited by the range of material available. Sometimes new evidence does not fit the existing pattern, but rather than reject it, we may have to reassess earlier views.

The process by which these changes in living organisms have occurred in geological time is termed **evolution**. The essential steps in the theory of evolution as put forward by Darwin are referred to in Chapter 8. Biologists accept that essentially the same processes have been responsible for evolutionary changes in all living organisms, including those that have led to the emergence of human-like animals and ultimately of *Homo sapiens*, today's human species. Some people, usually because of conflicts with religious beliefs, find it difficult to accept the principle of evolutionary change, particularly when applied to humans. For example **creationists** believe that each living species was created by a supernatural being rather than by organic evolution, though there is overwhelming biological evidence which supports the facts of human evolution. Acceptance of evolution by change from common ancestors is central to our understanding of human evolution. Table 10.1 depicts recent events associated with human (hominid) evolution.

A **classification** of living organisms can help us see the relationships between groups and so trace their evolutionary history. In a *biological* classification, organisms are grouped together if they show similarities and they are put in separate groups if they show differences. Following on from the Darwinian theory of evolution, in a classification scheme we should expect to find patterns of close (or distant) relationships which reflect the evolutionary history of the organisms. The term **phylogeny** describes the way we attempt to reconstruct the evolutionary history of any group by means of features and relationships used in a classification.

Humans as primates – our nearest living relatives

Humans belong to the classification group or Order known as the Primates. Table 10.2 gives an outline classification of some living primates. Through this

DEFINITIONS

Palaeontology is the study of biological events in the distant past, including the study of fossils and their evolutionary relationships. When possible, it includes study of the ecology of fossils. **Archaeology** is the study of recent events in human history.

BACKGROUND

Timescale of events associated with evolution

In attempting to reconstruct the past we need to establish a timescale, though, inevitably, some events are dated only approximately. For early **geological** events, we are dealing with millions of years and it is often hard to visualise such enormous spans of time. But for the more recent **biological** events, particularly when human-like animals appear, we can begin to have a more realistic feel of the time dimension. Assuming a human generation spans about 25 years, going back 100 years in a family tree would give about four sets of parents (or child to great grandparents). Extending further backwards suggests there were about 400 generations between us and the beginning of the Neolithic (10 000 years ago), and about 80 000 generations to the earliest known evidence of humans as *Homo habilis* (approximately 2 million years ago).

One stage of hominid (human) evolution merges into another and overlaps with the next, but it is convenient to associate the main events with the successive geological epochs. The names of the periods and epochs need not be remembered, but are given to help relate the biological events in the evolution of living organisms to the geological events.

Table 10.1 *Geological timescale and events in human evolution. The times in millions of years (Myr) indicate the approximate beginning of each period of epoch.*

Period	Myr	Epoch	Events
Quaternary	0.01	Holocene	modern *Homo sapiens*; Neolithic culture
	2	Pleistocene	*Homo erectus*, Neanderthal man, *H. sapiens*; Palaeolithic cultures
Tertiary	5	Pliocene	australopithecines, *Homo habilis*
	23	Miocene	hominoids, origin of hominids
	35	Oligocene	anthropoids, origin of hominoids
	56	Eocene	early Old World simians
	65	Palaeocene	prosimians

classification, we can trace the phylogenetic relationships between groups of primates. The classification also reflects trends and changes that have taken place in their evolutionary history. The underlying assumption is that similarities between groups indicate a closeness of relationships and the degree of difference is a measure of how long ago the groups diverged.

Apes and humans – You can see (in Table 10.2) that humans are in the **hominid** family (Hominidae). The closest relatives of the hominids are the great ape family, which includes orang-utans, gorillas and chimpanzees. Gibbons are placed in a separate family. These three families are grouped into a 'superfamily', the Hominoidea, which includes apes and humans.

Apes and humans are distinguished from the rest of the primates by being tailless and by having the ability to swing their arms freely, around and up and over their heads. Gibbons inhabit the forests of Thailand and other parts of south-east Asia. They show a characteristic graceful arm-swinging locomotion, a habit described as **brachiation**, in which they swing adeptly from branch to branch through the trees, one arm alternating with the other and hanging from the branches as they go. On the ground they walk or run a little uneasily on their two hind legs (bipedally), often waving their arms in the air. Their long

HUMAN EVOLUTION

Table 10.2 *Some living primates and how they are classified. [Note that the scientific (latinised) names are used to give a complete picture, but in most cases you will find that the common names can be used.]*

Suborder	Infraorder	Superfamily	Family	Common name
Prosimii (prosimians)	Lemuriformes (lemuriforms)	Lemuroidea (lemurs)	Lemuridae	lemurs
Anthropoidea (simians or anthropoids)	Platyrrhini (New World simians)	Ceboidea (New World monkeys)	Cebidae (true monkeys)	capuchins
				spider monkeys
			Callitrichidae	marmosets tamarins
	Catarrhini (Old World simians)	Cercopithecoidea (Old World monkeys)	Cercopithecidae	cheek-pouched monkeys baboons
				leaf monkeys
		Hominoidea (apes and humans)	Hylobatidae	gibbons
			Pongidae (great apes)	orang-utans gorillas chimpanzees
			Hominidae (hominids)	humans

Look first at the **families** (in the highlighted box) to see how these are grouped together. In the text we work from the familiar (humans), to the less familiar (lemurs). Use the table to see the underlying basis for the major divisions in the classification.

slender fingers, with delicate nails, are remarkably human-like as they explore and pull apart potential foods.

The great apes (chimpanzee, orang-utan and gorilla) are progressively heavier (male gorillas can weigh up to 200 kg) and, compared with gibbons, move less freely through the trees. On the ground, they walk on all fours (quadrupedally), rolling their fingers into the palm of the hand and using their knuckles for support. When upright, they waddle rather than walk. Chimpanzees seem very human-like in some of their behaviour patterns and facial expressions, so it is not difficult to appreciate that apes and humans are closely related. Orang-utans are found in south-east Asia; chimpanzees and gorillas inhabit parts of central Africa.

Old World monkeys – Moving up the table we find the Old World monkeys, such as leaf monkeys, baboons and cheek-pouched monkeys. The Old World monkeys inhabit Africa and Asia. They have tails, and walk on all fours on the ground or in the trees, using the palm of the hand for support. Leaf monkeys feed almost exclusively on leaves (hence their name) and other Old World monkeys are mainly fruit eaters. The Old World monkeys have not developed the free arm-swinging movements of apes and humans (Hominoidea).

New World monkeys – The New World monkeys are found in Central and South America. This is a fairly diverse group, whose features make them a little hard to classify satisfactorily. Marmosets and tamarins are small (the pygmy marmoset weighs only 100 g) and they have claws rather than nails. An unusual

feature is their diet, which often consists of gums scraped from the surface of trees. Spider monkeys are larger and noted for their long prehensile tail, which is used for grasping branches and thus helps with locomotion. Spider monkeys feed mainly on fruit and leaves. Another group of New World monkeys includes capuchins, squirrel monkeys and owl monkeys. These are larger than marmosets and tamarins, and some use their tails in a prehensile way for grasping.

Lemurs – Present-day lemurs live only on the island of Madagascar. The several existing species are clearly relics of a formerly more widespread and primitive family. Fossil representatives have been found in North America and Europe. In their history, the group appears to have been overtaken by the more successful monkeys and apes, but just a few species have probably survived in Madagascar because here they were isolated from competitors. Lemurs run around on the branches of trees and are herbivorous. They have tails that are not prehensile. They show a certain amount of social grouping but less manual dexterity than monkeys or apes. In a number of features, they show themselves to be the most primitive of the primates.

Figures 10.1 and 10.2 show examples of living primates, and their skeletons and movement. As a group, primates show adaptations to an arboreal way of life (living in trees) and are mainly vegetarians. Associated with this lifestyle is well-developed stereoscopic vision, which is essential for monkeys and apes for accurate judgement of distance when they leap rapidly through the trees. Looking at primates as a whole, this classification allows us to see trends within the order, from lemurs and monkeys, which are representatives of earlier and more primitive forms, through to hominoids (apes and humans), which are considered to be more recent and more advanced types. Similarities and gradual changes observed between different families give evidence of shared ancestry, but not necessarily of direct descent.

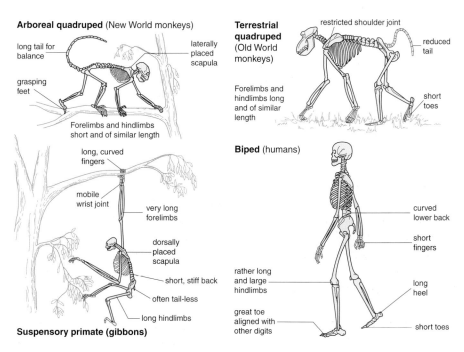

Arboreal quadruped (New World monkeys)

long tail for balance

laterally placed scapula

grasping feet

Forelimbs and hindlimbs short and of similar length

Terrestrial quadruped (Old World monkeys)

restricted shoulder joint

reduced tail

Forelimbs and hindlimbs long and of similar length

short toes

long, curved fingers

mobile wrist joint

very long forelimbs

dorsally placed scapula

short, stiff back

often tail-less

long hindlimbs

Suspensory primate (gibbons)

Biped (humans)

curved lower back

short fingers

rather long and large hindlimbs

long heel

great toe aligned with other digits

short toes

Figure 10.2 *Skeletons and movement in living primates. These primate skeletons show progression from an arboreal quadruped through to humans as upright walking bipeds.*

Figure 10.1 *Examples of living primates – different primate families (from top to bottom): ring-tailed lemur; long-haired spider monkey; olive baboon; male silverback mountain gorilla*

Questions

Reorganise the classification chart of the primates (Table 10.2) into a family tree, from suborder through to family. Include in it the features which can be used to separate the groups.
Then add the features to do with skeletons, skulls and teeth to the information in the family tree you devised for the primates.

• Use some of these features and devise a dichotomous key which could be used to identify members of the main groups within the primates.

• List the features which characterise and distinguish humans from other living primates.

Evidence for human evolution

We can use evidence from different sources to piece together the events that have led to the evolution of the human species and use this evidence to trace the probable phylogenetic links between related organisms, past and present. In this chapter we look at evidence from

• comparative anatomy (including skeletons and skulls)
• fossils and geochronology
• biochemical and molecular sources (including immunological studies, amino acid sequences and DNA)

Comparative anatomy – comparing humans with other living primates

This section takes a closer look at living hominoids (apes and humans) and in particular compares their skeletons, including limbs, skulls and teeth. It focuses on the stages in 'getting upright' to become bipedal and increased development of manipulative skills with the hands. This comparison highlights trends in living hominoids. This then helps you to interpret similar trends found in fossils of early human species as discussed later in this chapter (see pages 144 to 145) and which provide evidence for evolutionary changes in early humans.

Quadripedal to bipedal walking – the descriptions on pages 129 and 131 indicate that Old World monkeys walk on all fours and are clearly quadrupeds. The apes, however, including gibbons and the great apes (chimpanzees, orang-utans and gorillas), show features of both quadrupeds and bipeds (walking upright on two legs). In evolutionary terms, the apes represent an important stage between quadrupedal and bipedal locomotion. Humans, as we know, are truly bipedal. A comparison made of relevant features in chimpanzees and humans is shown in Figures 10.3, 10.4 and 10.5 and

Figure 10.3 Getting upright and shifting the balance: comparing chimpanzees and humans

(1) In chimpanzees, the centre of gravity (centre of mass) lies above a line between the four legs, so when the chimpanzee stands on two legs, the tendency is to tip forward to the more stable position on four legs.
(2) In humans, the position of the centre of mass is in the pelvic region and when standing upright acts downwards along the line shown, down through the legs to the feet.

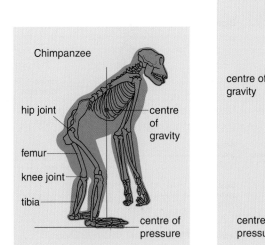
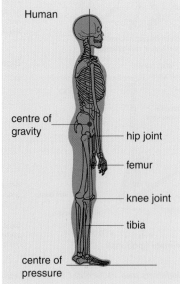

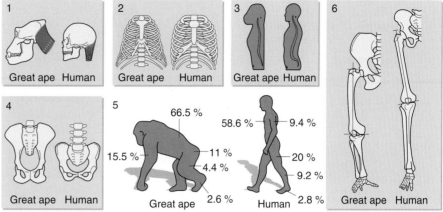

Figure 10.4 Getting upright and walking on two legs: comparing great apes and humans

(1) The human head is well balanced on top of the vertebral column; a great ape's head is held in position by powerful neck muscles.

(2) The human ribcage has become more barrel-shaped than that of great apes, probably because human arms are not used for locomotion.

(3) The human vertebral column has additional forward curves in the neck and lower back. These reverse curves help to bring the head and trunk above the centre of gravity (centre of mass) in the upright position.

(4) The human pelvis is broader and lower than that of the great apes.

(5) Human legs are longer than their arms so a greater proportion of the body weight (hence their centre of gravity) is lower. [The percentages indicate the proportion of the body weight in that region.]

(6) The human femur is angled outwards towards the knee, allowing the knee to be brought under the body and close to the line of action of weight. Humans extend the leg fully when walking and the bones of the leg form a straight line.

Skeletons, locomotion and posture – features and trends

- **Limbs** – quadrupeds have limbs about the same length, short if the quadruped is arboreal, long if terrestrial; apes have proportionately longer arms; humans have relatively longer legs.
- **Vertebral column** – the trend is from an arched vertebral column (Old World monkey) to a straight vertebral column (chimpanzee) and to the human vertebral column with reverse curvatures in lumbar (lower back) and cervical (neck) regions.
- **Centre of gravity** – S-shaped vertebral column of humans brings head and trunk above centre of gravity (centre of mass); body weight (mostly in vertebral column and bones of legs) acts vertically through centre of gravity at hip joints.
- **Pelvis and shoulder girdle** – (apes) for walking and climbing on all fours + swinging in branches whereas in humans short wide pelvis allows bipedal walking and the broad, flexible shoulder blade allows free arm rotation.
- **Big toe** – opposable in apes but non-opposable in humans (note that humans cannot grasp with the big toe – this feature gives a characteristic human footprint); human has an arch.
- **Knee** – leg bones straighten when walking upright (femur in straight line with lower leg bones).
- **Ribcage** – barrel-shaped in humans (no longer use arms in locomotion).

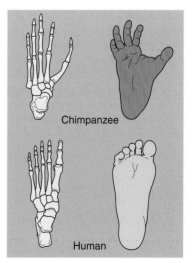

Figure 10.5 Comparing feet: chimpanzees and humans

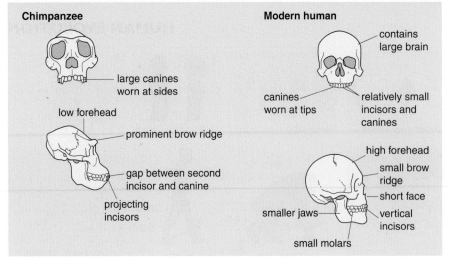

Figure 10.6 Skull and jaws: comparing chimpanzees and humans

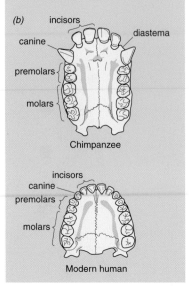

Figure 10.7 Changes in teeth and jaw shape: comparing chimpanzees and humans. (a) lower jaw. (b) surface view of the upper jaw in chimpanzees and humans.

Skulls and brain size, jaws and teeth – features and trends

- **Position of skull when upright** – relationship with joints of neck (centre of gravity alters), position of foramen magnum (hole through which spinal cord goes) is further back in apes, eyes at front.
- **Size of skull (in relation to body size)** – enlarged cranium accommodates increased brain size. Humans have the largest brain volume in relation to body weight.
- **Shape of skull** – loss of prominence of brow ridges; protruding jaw to a more shortened face with rounded and less protruding jaw.
- **Jaw features** – V-shaped jaw to more U-shaped.
- **Teeth modifications** – loss of large conical canines, reduced molars.

Hands – features and trends

- **Prehensile** – able to grasp objects and pick up objects between thumb base and finger tips.
- **Opposable thumb** – can rotate so that fleshy tip can touch other fingers, giving highly mobile thumb joint (from Old World simians only).
- **Tactile sensitivity** – on palms, explore and hold objects for closer examination by eye.
- **Ridged patterns (fingerprints)** – convey sensory information to the brain.
- **Rotation at wrist** – proximal end of radius can rotate, hence forearm twists. This enables the hand to be flat on ground or turned upwards.
- **Length of fingers** – relatively short fingers in gorilla allow knuckle walking (cf. gibbons and humans).
- **Length of thumb** – relatively short thumb in gibbons does not interfere with arm-swinging locomotion (brachiation) in trees
- **Manipulative skills** – extreme versatility in manipulative skills and precision movements, e.g. precision grip, power grip; flexion (bending) of fingers at joints, rotation of fingers (especially thumb); mobility of hand (rotation at wrist).

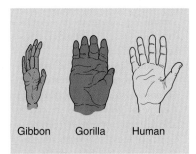

Figure 10.8 Thumbs and fingers: comparing gibbons, gorillas and humans. Note how in gibbons the relatively short thumb does not interfere with arm-swinging locomotion (brachiation) in trees. Gorillas have relatively short fingers (compared with gibbons and humans) and their fingers curve towards the palm, allowing them to use their knuckles for support when walking.

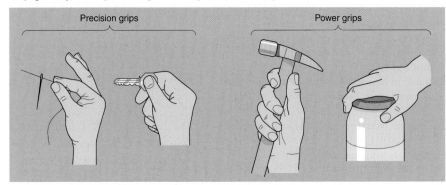

Figure 10.9 Human manipulative skills. The ability of the thumb to rotate and the mobility of the fingers allow humans to show enormous versatility in their hands and to carry out many precision movements. Some of the commonly used grips are illustrated in these diagrams.

summarised in the bulleted list 'Skeletons, locomotion and posture'. These features show the various changes that need to occur in the transition from quadripedal walking to bipedal walking.

Skulls, jaws and teeth – to walk upright, there need to be changes in the limbs and shifting the centre of gravity. In addition, comparison of the skull, jaws and teeth of chimpanzees and humans shows features that need to change in the adjustment to upright walking and a general increase in size of the brain in relation to the size of the body. This larger brain size gives the potential for increased brain capacity and hence of greater complexity in behaviour patterns. This comparison is made in Figures 10.6 and 10.7 and summarised in the bulleted list 'Skulls and brain size, jaws and teeth'.

Hands and manipulative skills – still linked with the shift to walking upright, we can see how comparison of features of the hands shows that human hands have more mobility and a far greater range of activities that they can carry out, when compared with gibbons and gorillas. This comparison is made in Figure 10.8 and summarised in the bulleted list 'Hands'. Some of the specialised manipulative skills of the human hand are illustrated in Figure 10.9.

Fossils and geochronology

The preserved remains of living organisms are usually described as **fossils** when they are more than about 10 000 years old. In plants, the fossilised material is likely to be from the wood, veins of leaves, pollen grains or seed coats, and in animals mainly from bones and teeth. Soft parts are usually not preserved. The fossils may be the actual remains of the living material, or may be in the form of imprints, moulds or casts of the former living material. The hard parts of the material may have undergone changes. For example, protein in bones may be replaced by minerals containing calcium or silica in a process known as **mineralisation**. This makes the fossil harder and stronger than the original material. In some cases, the original material is completely replaced during these mineralisation processes, leaving a cast or mould of the original shape.

Any fossils that are discovered are likely to represent only small fragments of the original living organism. After recording where they are found, the fossils are then compared with other relevant material and attempts made to understand the significance of the discovery in relation to other known fossils or organisms. One important step is to estimate the age of the fossil and thus establish the sequence of events and changes that took place in the evolution of the particular group of organisms.

(a)

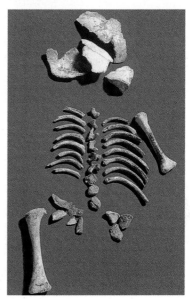

(b)

Figure 10.10 (a) Laetoli footprints, a trail of hominid footprints, fossilised in volcanic ash, found at Laetoli, Tanzania. The trail dates from over 3 million years ago and probably belongs to Australopithecus afarensis. *The footprints show a well-developed arch to the foot and no divergence of the big toe. They indicate upright walking and probably represent two adults and a child. (b) Fossil bones of the skeleton of a Neanderthal infant.*

QUESTION

Suggest conditions in which decay of dead organisms would be very slow.

BACKGROUND

Fossils – their formation and how they can give evidence of ancestral forms

Formation of fossils is a chancy business. When living organisms die, the processes of decay by microorganisms usually set in rapidly. In addition, the effects of weathering by heat, cold, wind and water (particularly when acid) make corpses of animals disappear quite fast. Parts of animal corpses might be consumed or dragged elsewhere by scavengers. Just occasionally, conditions at the time of death allowed the living material to be buried so that some parts of the original structure became preserved without undergoing decay.

Where fossils did form, it is usually because deposits (known as sediments) of silt, clay, sand or gravel settled over the remains of organisms and buried them. These particles often came from flowing rivers or may have built up in the bottom of lakes. Deposits of sand, silt or clay also accumulate in caves. Caves may have been places where carnivores dragged their prey and certainly early humans used caves for shelter or burial, and these accumulated materials would have been suitable for fossilisation. Sediments are usually soft when formed, but over a period of time they harden to form rocks.

Active volcanoes can contribute to the formation of fossils. Volcanic ash may fall over wide areas and become part of the soil. Close to the volcano, a thick layer of ash may cover the landscape and hot lava and mud, flowing out from the volcano, may trap remains of organisms underneath. Fossil footprints probably originated as footprints in soft mud, which were then rapidly filled in with another contrasting material, such as volcanic ash.

Determining the age of fossils – geochronology

The term **geochronology** refers to the methods used to find the age and sequence of geological events. When fossils are found, it is important to know the approximate age of the material. This may be determined in relative terms (compared with other material or events that occurred at about the same time) or it may be calculated on an absolute time scale, giving an approximate age in calendar years. It is important to have a knowledge of the approximate age of the material when comparing fossil material obtained from different places.

The techniques used (see Figure 10.11) depend on both the type of material and on its age. Often the dating is applied to material in the rocks immediately below and above the fossil rather than to the fossil itself.

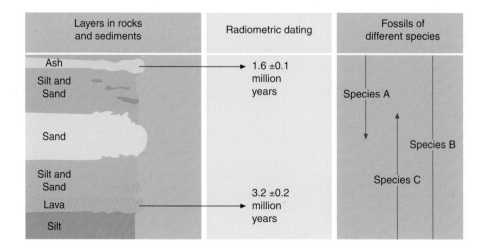

Figure 10.11 Ways of dating fossils by looking at the age and sequence of geological events

- looking at the **layers of sediments** that have built up around the fossil. The characteristics of the rocks and the sequence of the layers give clues about successive events over a period of time. Generally this approach can be used to establish a relative time sequence for the events that occurred and can give an indication as to the age of rocks (and fossils found in them) with similar patterns at different sites.
- using **radiometric techniques** with material to give an absolute measure of time.
- looking at **fossils of other species** associated or overlapping with the fossils being examined. The history of these other species may have been established from finds elsewhere. A good example is found in the horse, elephant and pig. The fossil records of these three species reveal a fairly complete picture of the stages of their evolution. When these remains are found alongside human fossils, they can help give a relative date to the associated human material.

Radiometric dating methods are based on measurements of rates of radioactive decay. Radioactive forms (isotopes) of elements are assumed to decay at predictable rates, regardless of surrounding conditions. The 'half-life' indicates the rate at which the radioactive element is known to decay. The proportion of the radioactive isotope in a sample of material (compared to the normal form of the element), can give an absolute measure of the time at which the material originated. The two methods most useful in the study of human fossils are **potassium–argon** dating and **radiocarbon** dating.

Potassium–argon dating is used for dating material in volcanic rocks, older than 1 million years. Lava from volcanoes releases radioactive potassium (^{40}K). This is incorporated into deposits of sediments that form around the volcano. Any argon present at that time is driven off. This effectively sets the 'time clock' to zero. In material that is examined later, any argon present is assumed to have come from the original radioactive potassium. In a specimen

- the ratio of ^{40}K to ^{40}Ar gives a measure of the age of the specimen
- the half-life of ^{40}K is used to calculate the actual age

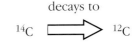

decays to

^{40}K ⟹ ^{40}Ar

potassium   argon
(radioactive   (stable isotope)
isotope)

Radiocarbon dating is used for dating organic material, such as bone or shells, up to about 50 000 years old. Carbon exists mainly as the stable isotope (^{12}C) but includes a small proportion of the radioactive isotope (^{14}C). Plants carry out photosynthesis and take in carbon dioxide from the surrounding atmosphere. Some of this plant material is eaten and the carbon present becomes incorporated into the bodies of animals.

decays to

^{14}C ⟹ ^{12}C

carbon   carbon
(radioactive   (stable isotope)
isotope)

The radiocarbon dating method assumes that plants (growing many years ago) took in carbon dioxide with a ratio of ^{14}C to ^{12}C that was the same as that in the atmosphere of the time. When the living organisms died, the radioactive isotope (^{14}C) started to decay so that, over a period of time, the proportion of radioactive ^{14}C in any fossilised material changes. In a specimen

- the ratio of ^{14}C to ^{12}C gives a measure of the age of the specimen
- the half-life of ^{14}C is used to calculate the actual age

HUMAN EVOLUTION

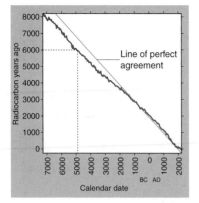

Figure 10.12 Radiocarbon dating estimates age by measuring the proportion of ^{14}C to ^{12}C in a sample of material. The graph shows radiocarbon dates in a sample of wood compared with age deduced from counting tree rings. A radiocarbon age of 6000 years corresponds to a calendar date of 4900 BC.

In tracing human evolution, fossils have provided important evidence. Fossil skeletons have given a definite record of the size of the individual, its posture and the type of locomotion; fossil skulls show how the shape of the face and jaw have changed and how the brain has increased in size. Inevitably there are uncertainties surrounding the precise details of the evolution of human or of any other species and there is controversy and debate even amongst the experts who find and interpret the fossils. Dates are only approximate, partly because of limitations in the accuracy of methods used to estimate the date of the material and partly because finding a fossil in one location does not give information about how long the species might have existed, nor does it preclude the form being present elsewhere, but where no fossil has yet been found.

Associated with fossils there may be **artefacts** (objects made or used by humans), such as shaped stones which would have been used as tools. When associated with fossil skeletons, artefacts have become an important source of evidence for building up a picture of how early humans lived and of their cultural development. Artefacts include flaked stones and bones and later evidence indicates use of fire, followed by more sophisticated expression in terms of jewellery or cave art. Artefacts continue to reflect the history and achievements of people, right up to the present day, through buildings, jewellery, paintings or even rockets launched into space.

Molecular clues to human ancestry

Immunological studies of blood sera – the antigen / antibody reaction

The immune system acts as a defence system and helps to protect the body from any harm that might be caused by the presence of a foreign organism. The immune system recognises and reacts to foreign molecules. These foreign molecules are often proteins and may be introduced on the surface of an invading organism. The foreign molecule is known as the **antigen** and this stimulates the production (by certain lymphocytes in the blood) of **antibodies**. These antibodies act against the antigen to destroy it or at least neutralise its effect. Antibodies are proteins and are specific to the invading organism, so antibodies produced by one organism can act as an antigen in another. If antibodies produced in response to a protein (the antigen) are mixed with the original protein, these now react against each other and form a precipitate. This response has been used to investigate the closeness of relationships in a range of animals.

The sequence of steps to be taken is summarised in Figure 10.13. Note that the rabbit serum is described as 'sensitised' when it contains the antibodies produced in response to antigens in the original human serum. If this rabbit serum is mixed again with human serum, it now reacts strongly and this is described as a 100% reaction. If the rabbit serum is then mixed with serum from other species, varying degrees of reaction (seen as formation of precipitate) may occur. A high percentage reaction indicates the species are closely related (in this case, to humans) whereas a low percentage suggests greater divergence, both in time and in relatedness. Some results are shown in Table 10.3.

Table 10.3 *Relationships shown by immunological studies of blood sera – comparison of the antigen / antibody reaction of a number of mammal species*

Species tested	Per cent reaction
human	100
chimpanzee	95
gorilla	95
orang-utan	85
gibbon	82
baboon	73
spider monkey	60
ruffed lemur	35
dog	25
kangaroo	8

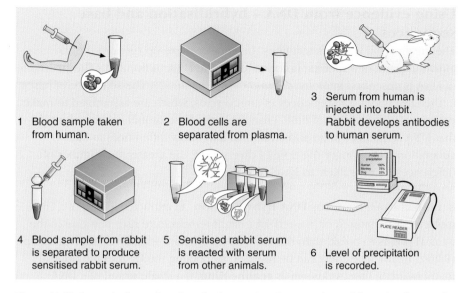

1 Blood sample taken from human.

2 Blood cells are separated from plasma.

3 Serum from human is injected into rabbit. Rabbit develops antibodies to human serum.

4 Blood sample from rabbit is separated to produce sensitised rabbit serum.

5 Sensitised rabbit serum is reacted with serum from other animals.

6 Level of precipitation is recorded.

Figure 10.13 Stages in the antigen / antibody reaction. Serum is derived from the plasma of the blood. It contains proteins but those involved in the clotting reaction (such as fibrinogen) have been removed

Amino acid sequences in proteins

Automated biochemical techniques can be used to work out sequences of amino acids in proteins or specific parts of them. Comparisons have then been made of equivalent parts of cytochrome *c* and of the α- and β-chains of haemoglobin from different species. In the haemoglobin molecule, the α-chain is composed of 141 amino acid residues and the β-chain of 146 residues. Humans and chimpanzees are identical in these two chains; orang-utans differ from humans in two amino acids and gorillas differ in four. This suggests that humans are more closely related to chimpanzees than they are to orang-utans or gorillas (Figure 10.14).

QUESTION

How far do these data for the antigen / antibody reaction (Table 10.3) match the relationships suggested by the classification of primates, given in Table 10.2?

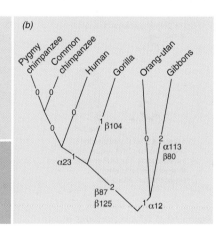

(a)

	β80	β87	β104	β125	α12	α23	α113
Human	N	T	R	P	A	E	L
Common chimpanzee	N	T	R	P	A	E	L
Pygmy chimpanzee	N	T	R	P	A	E	L
Gorilla	N	T	K	P	A	D	L
Orang-utan	N	K	R	Q	T	D	L
Gibbon	D	K	R	Q	T	D	H
Old World monkeys	N	Q	K	Q	A	E	L
New World monkeys	N	Q	R	Q	A	D	H
Tarsier	N	K	R	Q	A	D	H
Lorises	N	K	R	Q	A	D	H
Lemurs	N	Q	T	A	T	E	H

Key to acid residues:

A = alanine; D = asparagine; E = glutamic acid; H = histidine; K = lysine; L = leucine; N = aspartic acid; P = proline; Q = glutamine; R = arginine; T = threonine

Figure 10.14 Comparing amino acid sequences in chains of α-haemoglobin and β-haemoglobin in some living primates: (a) in the primates listed there are only seven positions in the chains which show amino acid differences; (b) possible evolutionary tree for apes and humans, based on these amino acid differences in the haemoglobins. The numbers within the branch lines indicate the number of amino acid residues that are different and the numbers beside the lines show the positions at which the differences occur.

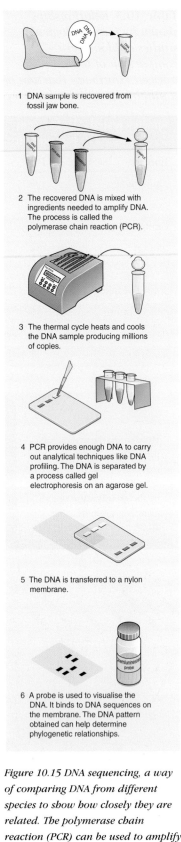

1 DNA sample is recovered from fossil jaw bone.

2 The recovered DNA is mixed with ingredients needed to amplify DNA. The process is called the polymerase chain reaction (PCR).

3 The thermal cycle heats and cools the DNA sample producing millions of copies.

4 PCR provides enough DNA to carry out analytical techniques like DNA profiling. The DNA is separated by a process called gel electrophoresis on an agarose gel.

5 The DNA is transferred to a nylon membrane.

6 A probe is used to visualise the DNA. It binds to DNA sequences on the membrane. The DNA pattern obtained can help determine phylogenetic relationships.

Figure 10.15 DNA sequencing, a way of comparing DNA from different species to show how closely they are related. The polymerase chain reaction (PCR) can be used to amplify very small amounts of material obtained from fossils.

Using evidence from DNA – hybridisation and base sequences

The information which determines the genetic make-up (genome) of an individual, and of a species, is contained in the DNA molecules. (The structure of DNA is described in *Molecules and Cells*, Chapter 2.) The sequence of bases in the DNA codes for sequences of amino acids, which are assembled to make proteins. Over a period of time, changes in the base sequence may occur, so that DNA sequences in an ancestral form may show differences in later generations. The longer the time of divergence, the greater the number of differences that are likely to have accumulated.

In the technique known as **DNA hybridisation**, an extracted length of a DNA molecule is denatured or 'melted' by a slight temperature rise. This causes the weak H-bonds to break, then the two complementary strands separate or unzip (Figure 10.16). The effect can be reversed by cooling the DNA to the original temperature, a process known as **reannealing**. If a single strand of a DNA molecule is allowed to reanneal with the complementary strand from another species, a hybrid DNA molecule is formed. The stability of this hybrid molecule reflects how closely the two strands were able to match up their complementary bases. The greater the number of bases that paired, the higher the temperature required to separate (melt) the hybrid molecule. DNA–DNA hybridisation shows the closest relationship to be between humans and chimpanzees, then between humans and gorillas and orang-utans. Gibbons and Old World monkeys are more distant.

The actual sequence of bases along certain lengths of the DNA molecule can be revealed by the technique known as **DNA sequencing**. The DNA is extracted, then cut into lengths using enzymes (see Figure 10.15 and *Tools, Techniques and Assessment*, Chapter 5) known as **restriction enzymes**. After electrophoresis, the resulting patterns of banding of the DNA fragments can be revealed by treatment with radioactive labels or dyes. This allows the sequence to be 'read' and different genomes can be compared. Studies of DNA sequences confirm that the genomes of humans are very similar to those of chimpanzees and gorillas, with greater differences from orang-utans and even more from gibbons.

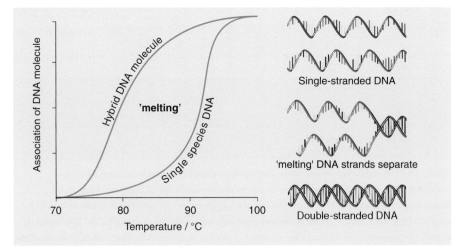

Figure 10.16 DNA hybridisation, a technique used to compare the genetic relatedness of DNA from different species. [PCR techniques are described in Chapter 9.]

Fragments of DNA can sometimes be recovered from fossil material, such as part of a jaw bone. If present in very small amounts, it may be necessary to amplify this material (make many copies) by means of the polymerase chain reaction (PCR). This then provides enough material to carry out the fingerprinting procedure or to match the DNA sequence with other known samples and thus provide evidence at the molecular level for phylogenetic relationships of extinct species.

QUESTION

How far do these relationships worked out from molecular evidence agree with the classification scheme given in Table 10.2?

Hominoid evolution

Our evidence for hominoid evolution (and divergence from apes), comes largely from fossils, even though relatively few fossils of early hominids have been found. Those that have been discovered are often fragmentary and rarely complete. Any new evidence may lead to other hominid species being proposed and it is likely that the designation of species will continue to change or be added to. Reference to Table 10.1 shows the likely timescale of events associated with the evolution of hominids.

Australopithecines – The australopithecines lived in Africa, between 4 million and 1 million years ago (4 to 1 Myr). They are extinct primates, but were undoubtedly related to early human ancestors. They have been described as 'man-apes' and also as 'ape-men' and clearly show features of both groups. [(The name *Australopithecus* is derived from the words 'southern' (*australo*) and 'ape' (*pithekos*, Greek word for ape.)] It is likely that they were a group of early hominids, sharing ancestry and overlapping with early human species, but not necessarily their direct ancestors. (Features of Australopithecines are summarised on pages 142 to 145.)

***Homo* species** – Fossil evidence for *H. habilis*, found only in Africa, is particularly scanty. There is general agreement that *H. habilis* represents the earliest known form that can be called human and that it originated in Africa. The later *H. erectus* also developed in Africa and from there migrated to Asia. The oldest dated *H. erectus* fossils come from Africa but later ones have been found in Asia. *H. sapiens* diversified into a range of forms, known as archaic, early modern through to modern. These spread from Africa and Asia into Europe, and eventually, in modern times, into most parts of the globe. Neanderthal man is known mainly from Europe, with an eastern distribution as far as central Asia.

DEFINITIONS

H. sapiens sometimes includes the Neanderthals, treated as a subspecies of *H. sapiens* (***H. sapiens neanderthalensis***), though some people consider the Neanderthals to be a separate species (***H. neanderthalensis***). The name is derived from the Neanderthal river in Germany where the first fossils were found. In this chapter, we will use the term 'Neanderthal man' and avoid designation of species or subspecies.

We can see distinct evolutionary trends in the 'species' of the genus *Homo*, from the early *H. habilis* through *H. erectus* to Neanderthal man and the later *H. sapiens*, but the progression does not necessarily mean that one form evolved directly from the preceding type. Certainly, *H. habilis* lived alongside the australopithecines, between 2 and 1.5 million years ago, before the australopithecines became extinct. *H. erectus* (from about 1.5 million years ago) also overlapped with the later australopithecines. *H. erectus* was probably a common ancestor to both modern *H. sapiens* and to Neanderthal man (Figure 10.22). Some people would support very much earlier dates for the origin of hominids, including the australopithecines and members of the genus *Homo*. (Features of *Homo* species are summarised on pages 144 to 145).

Climatic changes and the evolution of australopithecines and *Homo* species – About 4 million years ago, because of the Ice Age, a lot of water had been locked up as ice and rainfall had dwindled. Forest areas had shrunk and there were larger areas of tropical grasslands, known as savannah. There were relatively few trees and shrubs on these grasslands. The primates of that time were mainly tree-dwellers so, for them, the spread of grassland would have been a disadvantage. However, the australopithecines may have been able to take advantage of the loss of forest and changes to grassland. Footprints of australopithecines suggest they were bipedal in their walking and this would leave their hands free for making and using tools. By using even simple tools, such as stones or bones, they would have been less dependent on canine teeth for performing some tasks. Group cooperation might have helped survival on open grassland. Establishment of a home base would have allowed them to share food and help encourage social ties. With a relatively stable home base, there would have been the opportunity for the development of longer infant and childhood phases, with prolonged parental care.

Critical developments in the evolution of *Homo* species also coincided with considerable variations in climate – cold alternating with warm due to successive Ice Ages. Fluctuating temperatures led to changing sea levels: during a glacial period, sea levels were lower and there were land connections between some of the major continents (as we know them today). These land bridges allowed migration of evolving forms from one area to colonise another. During an interglacial period, when the ice melted, sea levels rose and some parts of the land masses became separated. Some branches of the *Homo* family tree may thus have become isolated on the different land masses and evolved along slightly differing lines (see Table 10.4, page 146).

Features and evolution of the australopithecines and early *Homo* species

In tracing the evolution of hominoids from their extinct ancestors to human forms as we know today, we can see a number of trends, some of which are summarised below:
- becoming upright
- development of bipedalism
- increase in manipulative skills of the hands
- increase in brain size
- development of language, together with other aspects of complex behaviour and intellectual activity associated with modern day humans

Similar trends are found in present day living primates and you may find it helpful to review the comparisons made on pages 132 to 135, illustrated in Figures 10.3 to 10.8, together with the lists of features and trends.

Fossil evidence, such as that of the Laetoli footprints (Figure 10.10) and that of the skeleton of the australopithecine named Lucy (Figure 10.18), provide support for the view that the early humanoid ancestors went through equivalent stages in the development of bipedalism. Development of bipedalism, and of hands that allow greater mobility of the thumb and fingers, are features that were enormously important in providing the potential for further evolutionary development. We can see the progression in the reconstruction of the life of early hominids and in the discussion of the cultural evolution and lifestyle of early humans.

Fossil skeletons assigned to the australopithecines have been found at different sites in Africa. They fall into two groups: a lightly built, or **gracile**, form and a more ruggedly built, or **robust**, form (Figure 10.20). The gracile form includes two species: *Australopithecus afarensis* and *A. africanus*. *A. afarensis* was found in the Afar region of Ethiopia, and the first partial skeleton found here was nicknamed 'Lucy'. Other *afarensis* fossils have been found at Laetoli in Tanzania, the site of the oldest known human footprints (Figure 10.10). The robust form includes *A. robustus* and *A. boisei*, though some workers feel these are sufficiently different from the gracile forms to justify a separate genus, *Paranthropus*. These fossil skeletons are far from complete, but enough can be pieced together to build up a picture of these primates when they were living.

Australopithecines show a number of features characteristic of humans in the genus *Homo*. Gracile australopithecines like Lucy probably walked upright, though perhaps a little awkwardly, and may have spent some time in the trees out of reach of predators (Figure 10.18). Lucy was about the same height and weight as a modern 6-year-old girl. The narrow hips would give a narrow birth canal in females, consistent with a relatively small skull size, which in turn indicates a relatively small brain. *A. afarensis* represents the oldest australopithecine that has been found and probably resembled a small, upright, dark and hairy chimpanzee. The robust australopithecines, *A. boisei* and *A. robustus* were larger and heavier, with larger brain size. In height and weight *A. robustus* approached that of a modern human, though the brain size was much smaller. The arms were relatively long, but the teeth show reduced incisors and canines, a feature associated with hominids.

The evolutionary trends and main features of the australopithecines and of *Homo habilis*, *H. erectus*, Neaderthal man and *H. sapiens* are summarised in Figures 10.18 to 10.21 (on page 144) and in the bulleted lists on page 145.

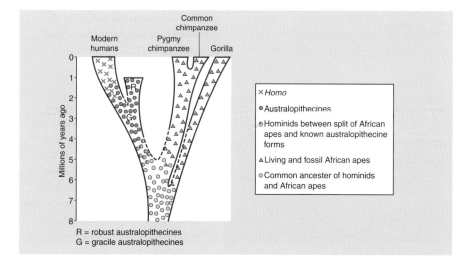

R = robust australopithecines
G = gracile australopithecines

Figure 10.17 The australopithecines are ape-like hominids, and probably branched away from the African apes between 4 and 5 million years ago

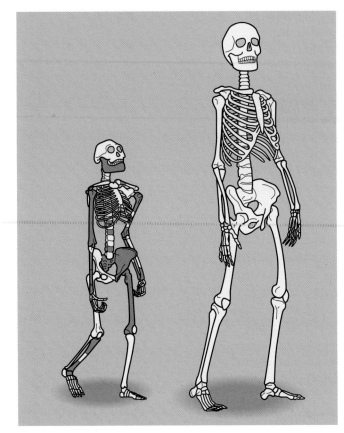

Figure 10.18 *Skeleton of the australopithecine nicknamed Lucy, seen alongside modern human skeleton. The shaded portions show actual fossil bones, the remaining parts have been reconstructed from other available evidence. Lucy was only 105 cm high, though other australopithecines were taller.*

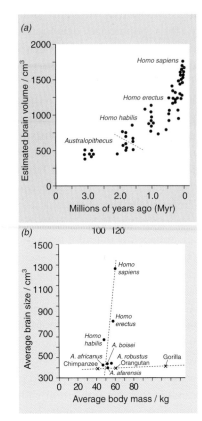

Figure 10.19 *Brain size in the australopithecines and the genus* Homo *compared with some modern primates: (a) increase in size of brain over time; (b) increase in size of brain in relation to body mass*

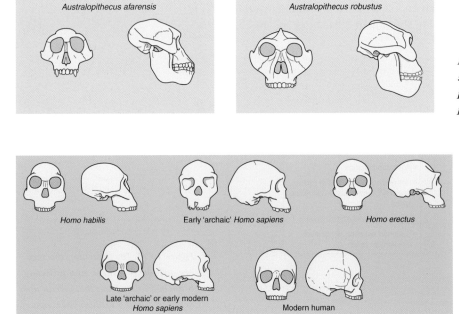

Figure 10.20 *Front and side view of fossil skulls of australopithecines. Note the presence of the low flat forehead and protruding jaw*

Figure 10.21 Homo *skull features: comparing* H. habilis, H. erectus *and* H. sapiens. *(Note the general trends from the earlier* H. habilis *to modern humans): prominent brow ridge of* H. habilis *recedes to almost no brow ridge in modern* H. sapiens; *size of skull increases, allowing development of larger brain; jaw becomes less protruding.*

Australopithecines and *Homo* – summary of features and trends

Australopithecines

Australopithecus afarensis – the **gracile** australopithecines, lived about 4 to 2.5 million years ago.

A. robustus – the **robust** australopithecines (or genus *Paranthropus*), lived about 2.6 to 1 million years ago.

- upright, height 1 to 1.5 m
- weight from 30 to 80 kg;
- relatively long arms;
- build
    - graciles: light build, some sexual dimorphism
    - robust forms: heavy build, some sexual dimorphism;
- brain size 400 to 530 cm^3;
- skull form
    - graciles: flat forehead (higher in *A. afarensis*), projecting face
    - robust forms: prominent crest on top of skull, flatter face;
- brow ridges – prominent in *A. afarensis*; also present in robust forms;
- jaw and teeth
    - graciles: incisors and canines relatively large in *A. afarensis*;
    - robust forms: very thick jaws, small incisors and canines, very large molars;
- location – eastern and southern Africa.

Figure 10.22 Skull of Neanderthal (left) compared with that of modern human (right), together with a 'model' or reconstruction of a Neanderthal face. This emphasises the Neanderthal flattened, broad shape, swept-back cheekbones, large front teeth, large nose and brow ridge above the eyes.

Homo habilis

(about 2.4 to 1.6 million years ago) small and large forms recognised;
- upright, height 1 to 1.5 m;
- weight about 50 kg;
- relatively long arms;
- pelvis allowed birth of babies with big heads;
- brain size 500 to 650 cm^3 (small), 600 to 800 cm^3 (large);
- cranium shape (see Figure 10.21);
- brow ridge, forehead slopes backwards;
- jaw light to robust but protrudes; small canines, narrow molars (small to large);
- location – Africa.

Homo erectus

(1.8 to 0.3 million years ago)

- height 1.3 to 1.5 m;
- weight about 40 to 75 kg;
- brain size 750 to 1250 cm^3;
- skull long, flat and thick;
- brow ridge, receding forehead;
- protruding jaw, teeth smaller than *H. habilis*;
- location – Africa, E and SE Asia, spread to Europe.

Neanderthal man

(150 000 to 30 000 years ago)

- height 1.5 to 1.7 m;
- weight about 70 kg;
- brain size 1200 to 1750 cm^3;
- skull thinner, large nose, protruding mid-face region;
- brow ridge reduced, low receding forehead;
- some development of chin, teeth similar to archaic *H. sapiens* but smaller;
- location – Europe and western Asia.

Homo sapiens

(archaic 400 000 to 100 000, early modern 130 000 to 60 000 years ago)
- height 1.6 to 1.85 m;
- weight about 70 kg;
- brain size 1100 to 1400 cm^3 (archaic), 1200 to 1700 cm^3 (early modern);
- skull higher, higher and more upright forehead, face less protruding;
- almost no brow ridge in early modern;
- smaller jaws, chin developed, teeth smaller (small molars, vertical incisors, small canines);
- location – Africa, Asia, Europe.

Table 10.4 *Climate and sea level changes while hominids were evolving, with possible stages of evolution in different parts of the world. (See pages 144 to 145 for details of the different hominid forms.) The 'tools' give an indication of the stage of cultural development, but evidence of actual tools used depends on the area and sites excavated.*

Epoch	Estimated years ago	Climatic conditions	Probable stage of hominid evolution	Tools
Middle Pleistocene	250 000–180 000	mainly temperate or cool, some glacial intervals • *large oceans, small ice caps*	• Europe – transitional 'archaic' *Homo sapiens*, early Neanderthals • Africa – late 'archaics' • China – late *H. erectus* and 'archaic' *H. sapiens*	Acheulian
	180 000–130 000	full glacial, some milder intervals • *small oceans, large ice caps*	• Europe – early Neanderthals, fixation of Neanderthal features • Africa – transitional to early Moderns • China – 'archaic' *H. sapiens*	
	130 000–115 000	last interglacial, warm conditions • *large oceans, small ice caps*	• Europe and Middle East – Neanderthals • Africa – first Moderns (and possibly Middle East) • China – 'archaic' *H. sapiens*	
	115 000–75 000	temperate / cool	• Middle East – Neanderthals and Moderns • China – 'archaic' *H. sapiens*	Mousterian
Late Pleistocene	75 000–30 000	cool / glacial • *ice caps increasing*	• Europe – Moderns appear; Neanderthals extinct • Far East – Moderns appear	microliths

Cultural evolution and lifestyle of early humans (Palaeolithic and Neolithic stages)

In tracing the main stages of the evolution of humans, we have described how we believe they adopted first a terrestrial rather than arboreal way of life, became truly bipedal and then developed the ability to use their hands. With this came enormous development of the capability of the brain.

Cultural development progressed alongside the physical changes described in this chapter. Evidence for cultural evolution is mostly indirect. Sticks and other wooden objects were probably used as tools, but it is unlikely that these would be preserved. The earliest objects recognisable as tools were stones, shaped and worked in different ways. Such stones have been found associated with fossil skeletons and skulls and other evidence of human existence.

In the Palaeolithic (Old Stone Age), from over 2 million years ago, there was considerable use of stone tools but no indication of settled agriculture. The Neolithic (New Stone Age), from about 10 000 years ago, was characterised by the domestication of plants and animals and the beginnings of true agriculture. These represent stages of cultural development that probably occurred at different times in different parts of the world. We shall never really know how early human communities interacted with each other, but we can speculate and build up at least a partial reconstruction of cultural life from about 2.5 million years ago.

> **QUESTION**
> • Draw up a chart to compare the australopithecines with the *Homo* species.
> • List features which characterise and distinguish modern humans from their probable ancestors. In particular, note when the human brain size and shape of the modern face became established.
> • Then look back to your earlier comparison of modern humans with other living primates. Compare them with descriptions given for the extinct australopithecines and for modern apes.

Palaeolithic (Old Stone Age)

This section describes the basic tools of the Palaeolithic stage, followed by a reconstruction of the lifestyle of these early humans, as they showed the early development of human lives including utilisation of tools and technology, and expression through art and religion.

The earliest stone tools were probably simple pebbles used for hammering, or knocked about to give a sharp edge, which was useful for more complex tasks. Shapes similar to those found at early hominid sites have been produced recently by workers simulating the activities of early hominids. If one stone is hammered with another (the 'core'), pieces of stone known as 'flakes' break away. The flakes have sharp edges and can be used for various purposes, such as cutting up meat or shaping a piece of wood for use as a tool. The term **knapping** is used for a person working stone in this way.

The earliest examples of stone tools are described as **Oldowan** (after the Olduvai Gorge site in Tanzania). Several types are recognised in the Oldowan 'toolkit', including a chopper and scraper (Figure 10.23). It is likely that *Homo habilis* used tools like these. More sophisticated tools in the form of hand axes have been found and are called **Acheulian** (named after St Acheul in France where they were first discovered). The earliest finds date from about 1.5 million years ago and tools from this toolkit were almost certainly used by *H. erectus*. These axes were elongated, with two worked faces, giving two cutting edges. A greater sense of purpose was needed to make them and this represented a considerable advance in the skills of the humans using them (Figure 10.24). Neanderthals used the **Mousterian** toolkit (named after Le Moustier in France where they were first found). These included a wide variety of tools, with finer stone flakes worked to provide a means of skinning, scraping, sawing or sharpening other tools (Figure 10.25). Gradually the tools became smaller, sharper and generally more specialised. The flakes were worked into blades to produce a diverse range of implements. These refined small stone tools, described as **microliths**, appeared towards the end of the Palaeolithic culture, about 25 000 years ago (Figure 10.26). They could have been used in spears, harpoons, sickles and many other implements, showing potential for use in a wide range of activities.

Homo habilis

The brain of *H. habilis* was larger than that of the australopithecines and markings inside the skull indicate development of a region known as 'Broca's area', suggesting the beginning of language ability. *H. habilis* is associated with Oldowan stone tools and the form of their hands and thumbs confirms rudimentary ability to use tools. The jaw and tooth pattern is consistent with their being fruit-eaters and they probably also scavenged meat. The form of the foot shows they walked upright, but their limbs would also have enabled them to climb trees. They are thought to have lived by river banks in more wooded habitats (whereas the nearby australopithecines lived in more open areas of savanna grassland). *H. habilis* could have used the open areas for food gathering, but retreating to the wooded areas for protection. There is evidence that *H. habilis* made simple shelters but could also take refuge in trees from large predators.

Trimmed flake used as scraper   Chopper

Figure 10.23 Early stone tools of the Oldowan toolkit – the heavier stones can be used as choppers (for cutting meat or cracking open bones). The flakes probably broke away while working the stones and would have been useful for various purposes, certainly for cutting meat off large animals. The chopper is about the length of an adult thumb.

Figure 10.24 The Acheulian toolkit includes a wider range of shaped stone tools. The typical axes were more difficult to make than the Oldowan chopper. The rounded end of the Acheulian axe could be held in the hand allowing the edge to be used for chopping and slicing. (Shown about one quarter actual size.)

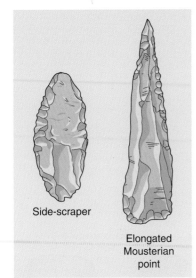

Figure 10.25 Tools of the Mousterian toolkit. (Shown about half actual size.)

Side-scraper

Elongated Mousterian point

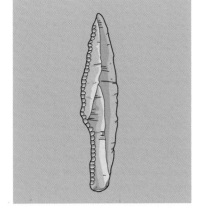

Figure 10.26 Example of a skilfully worked blade, associated with the Upper Palaeolithic cultures

This indicates the beginnings of a settled home base and is part of an evolutionary trend, noted earlier in the australopithicines (see page 142). With these more settled groups, there is the opportunity for sexual bonding to occur between males and females. This offers a security that in turn leads to greater care of their offspring, giving an extended childhood. In modern societies, we can see the advantages of how an extended childhood gives a longer period for development of behaviour patterns based on learning. The 'child', while still protected by parental care, can learn skills, acquire knowledge and develop social interactions with other members of the social group. This encourages cultural development, through the sharing of ideas and experiences, of the individual and of the society. This situation, familiar to us in modern societies, had its beginnings in the lifestyle of early hominids and is a feature that played an important part in the cultural development of the genus *Homo*.

Homo erectus

The larger brain size of *H. erectus* gave potential for more complex activities. There must have been some language ability and the position of the larynx indicates use of more elaborate sound patterns. Compared with *H. habilis*, *H. erectus* had a larger body size and was more muscular, which would have allowed mobility over wide areas, consistent with an active hunting lifestyle. They used tools from the Acheulian toolkit, but neurological control of the arm and hand was still restricted so that manipulations with the hand were limited. This is deduced from the size of the canal in the skull for nerves entering the brain. They certainly would have used other tools, made from wood. In some areas (such as Java) no stone tools have been found but locally fashioned bamboo probably served the same purpose.

H. erectus were nomadic hunter-gatherers, hunting large game such as boar, bison, deer, horse and rhinoceros. This was made possible by their more elaborate tools, which could have included spears or lassoes. The hunting required cooperation, probably between males, and the catch would have been brought back to the home base to be shared with females and the rest of the families in the group. Their teeth suggest a mixed diet; increasing consumption of meat was supplemented with locally gathered fruits and other plant material. Evidence of use of fire comes from charred remains and ash. They would have used fire for roasting meat and probably for keeping warm in some of the colder parts of their geographical range. They built fairly elaborate shelters or often used caves for their living places.

Neanderthal man

Neanderthal man was probably a side-branch in the evolutionary history of modern humans. They were short, stocky and muscular, well-built to withstand the colder climate of the Ice Age in Europe at that time. Their brain was larger than that of modern humans, though the Neanderthal skull was a little thicker, the jaw protruded and lacked a chin. The nasal cavity was enlarged – a feature that would have been an advantage in a cold climate by conserving heat and moisture from the freezing air outside. Their speech ability was probably still limited compared with *H. sapiens*. They used an elaborate range of tools from the Mousterian toolkit. They probably also used wooden tools or thongs, but these are unlikely to survive with other remains. The neck vertebrae show an enlarged hole, indicating that the spinal cord had become large enough to allow fine neurological control of manipulations with the hands. Their teeth were strong, and probably helped with some of the tasks they carried out.

Neanderthals lived in caves and also constructed large tents for protection. These would have been covered with skins from animals stretched over a framework of branches and weighted down with large bones (perhaps from mammoths). Skins were also used for simple clothing, helping to create a microclimate around the body, which is important for survival in the cold climate. They used fires for heat and for cooking. Neanderthals hunted big-game, capturing bison, horses, woolly cave bears and mammoths. This would have required cooperation within the group, suggesting rudimentary language skills. They may, for example, have used fire to drive herds over cliffs or into narrow gorges for mass slaughter. The bones of Neanderthal skeletons show frequent signs of wounds and damage, suggesting a fairly violent lifestyle. Nearer the home base, women and children would have gathered food such as berries, honey and smaller animals. Excess meat or fruit was perhaps conserved for another season by drying, smoking or freezing.

There are signs of community care and the beginnings of religious practices. Evidence for this is seen in burial sites where skeletons of old, arthritic or deformed individuals have been found. These people would not have been able to hunt, yet regrowth and healing of bone tissue suggests the injured were protected within the social group. Burial sites are recognised by the deliberate orientation of the skeletons, accompanied by remains of pollen grains, suggesting flowers had been buried with the corpse. The simplest form of art is seen in scratched pebbles, use of red colouring or a polished tooth. Neanderthals may have practised cannibalism, perhaps believing that they would gain strength from eating brains or bone marrow of their dead relatives and their spirits would carry on living. Marks seen on fossilised human bones have been interpreted as signifying cannibalistic practices, but may have arisen from activities of big animals or even falling rocks. This uncertainty reinforces the speculative nature of some of the interpretations made in reconstructing the lives of our ancestors. It is, however, clear that Neanderthals died out quite suddenly, and were replaced or overtaken by the modern form, *H. sapiens*.

Homo sapiens

With *H. sapiens* comes a rapid diversification and increase in complexity of the tasks which could be accomplished, consistent with greater mental capacity and improved motor control of activities, particularly with the hand. Cro-Magnon man, named after the site at Cro-Magnon in France, is taken to represent early modern *H. sapiens*, very close in ancestry to modern living humans.

H. sapiens, typified by Cro-Magnon man, is a (late) Stone Age human (upper Paleolithic), but is clearly identifiable as an immediate ancestor to modern human populations. The species probably originated in Africa rather than being direct descendants of Neanderthals, but Neanderthal man and *H. sapiens* overlapped for some time and there may have been some gene flow between the two populations. Compared with Neanderthals, Cro-Magnon man was of slighter build in body frame, though about the same height. Cro-Magnon man

> ### QUESTION
>
> How is the short, heavily built body of Neanderthals better able to survive in cold climates when compared with a tall lean body of present-day Masai people living in central Africa?

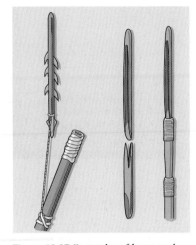

Figure 10.27 Examples of bone and similar tools used by Cro-Magnon man

shows considerable advances over Neanderthals in capabilities (with respect to tools employed), sensitivity as shown by artistic expression, exploitation of the environment and independence of it in terms of shelter and protection from extremes of climate.

Cro-Magnon man created and used tools showing far greater diversity than previous types (Figure 10.27). They made specialised blade tools, worked from flints, then used them to shape bone, antler and ivory. They used bone needles with eyes for sewing and barbed hooks for fishing. They made harpoons and spears which could be thrown. They could thus attack animals at a distance, perhaps making greater use of the brain to find ways of trapping and ensnaring them. Catching fish added yet more variety to their already protein-rich diet, which was supplemented with fruits and nuts gathered and stored. Remains of baskets suggest they were used to help collect these items from the surrounding land. At certain times they lived in caves, well chosen for their aspect (to trap sun) and view over the open valleys. They also built large and elaborate shelters, sometimes with stone for the floor, the roof being propped up with mammoth bones. These homes may have housed several families. They made simple fur suits from animal skins, a way of retaining body heat and providing protection from climatic extremes.

An outstanding advance associated with these upper Palaeolithic humans was in their artistic culture. They carved patterns and pictures on bones and antlers, made bracelets from ivory and necklaces from coloured beads, pierced teeth from larger animals or sea shells (Figure 10.28). Remains hoarded in burial areas suggest accumulation of such ornaments was an important part of their lives. Cave paintings often depict large mammals, including bison, aurochsen, deer, horses, mammoths and ibex. Examples of such cave art are found particularly in France and Spain.

Figure 10.28 Late Palaeolithic art on a reindeer antler

The hunting-gathering lifestyle of the late Palaeolithic people seems to have been reasonably successful in terms of survival. Food was mostly plentiful, but if an area became depleted, these nomadic people moved on to harvest a fresh patch of countryside. There were large and small game as well as fish, and berries, nuts and other fruits from the bushes. At some stage, large seeds from wild grasses were deliberately harvested and added to their mixed diet.

Mesolithic (Middle Stone Age)

There was a relatively short transition period, known as the **Mesolithic** (Middle Stone Age), which spanned the end of the Palaeolithic into the beginning of the Neolithic. This coincided with the retreat of the last Ice Age and, with the melting of the ice, there was a general increase of temperature and rising of sea levels. Animals migrated northwards and plants became re-established in the temperate zones across Europe through to Asia. As tools became more refined, they were applied to a wide range of tasks. There is evidence of people using baskets and simple pottery. Quite elaborate huts and shelters were constructed and the lifestyle was becoming more sedentary. The Mesolithic stage merges easily into the Neolithic, characterised by the beginnings of agriculture and the settled life that goes with it.

Neolithic (New Stone Age)

The Neolithic culture dates from around 10 000 years ago, though similar patterns of development of agriculture occurred independently, at different times, in at least three areas of the globe (Figure 10.29). The oldest was probably in the 'fertile crescent', an area in western Asia, bordered by the Taurus and Zagros mountains to the north, lying in the basin of the Tigris and Euphrates rivers.

Major centres of agricultural innovation: Plant and animal domestication apparently occurred independently and at different times in many different parts of the world. There were, however, three major centres of origin, whose influence spread geographically, eventually coming to dominate local innovation.

Meso America:
Maize, squash, beans, cotton, gourds. Llama, guinea-pig
[5000 years ago]

'Fertile Crescent':
Wheat, barley, emmer, einkorn, lentil, pea. Goats, sheep, cattle
[10 000 years ago]

China:
Rice, millet, soya bean, yam, taro, pea. Pigs
[7000 years ago]

Figure 10.29 There were at least three major centres for the origin of agriculture during the Neolithic cultural stage. Developments in the 'fertile crescent' about 10 000 years ago are described in the text. Two other important centres were China and Central America. There is evidence that settled agriculture existed in China about 7000 years ago, where the crops included rice, millet, soya beans, tea and yams. A later centre appeared about 5000 years ago, in Central America, where the crops were maize, beans, squash and peppers.

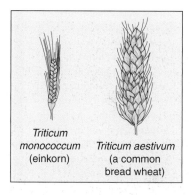

Triticum monococcum (einkorn)

Triticum aestivum (a common bread wheat)

*Figure 10.30 Wild wheat (*Triticum monococcum *= einkorn) compared with modern wheat (*T. aestivum*)*

Wild grasses found in the 'fertile crescent' area today are related to modern wheat and barley, and in the same area the wild animals include sheep and goats, which were among the earliest animals to become domesticated. This area also saw the development of the great early civilisations in Persia, Babylon and through to Egypt. Evidence of crops cultivated in this fertile crescent include wheat, barley, lentils and peas. Remains of cereal seeds show wheat similar to 'einkorn' and 'emmer' (*Triticum monococcum* and *T. turgidum* respectively), two species of *Triticum* believed to have given rise to modern varieties of wheat (*Triticum aestivum*) (Figure 10.30).

Artefacts from excavations show that Neolithic people had sickles with fine flint blades set into handles made from antlers (Figure 10.31). They used pestles and mortars for grinding their harvested cereal crops and had storage bins at their hut sites.

Wild animals that became domesticated include sheep (*Ovis orientalis* and *O. vignei*), goats (*Capra aegagus*) and aurochsen (*Bos primigenius*), the now extinct ancestor of modern cattle. These animals had a natural herding instinct, which predisposed them to successful domestication. Their ability to digest cellulose meant that they did not compete directly with humans for crops. Domestication of sheep is known from about 11 000 years ago, of goats from about 10 000. Aurochsen were first domesticated about 8500 years ago and wild boar (*Sus scrofa*) at about the same time. Supplies of meat and later milk were thus secured from one year to the next. Other animals that were domesticated and used for transport include the camel, donkey and horse, whilst cattle were valuable as a dual purpose animal (food and ploughing).

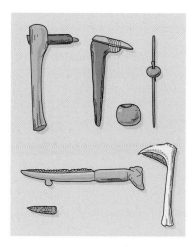

Figure 10.31 Some Neolithic agricultural tools

HUMAN EVOLUTION

Figure 10.32 Cattle used for ploughing in a rice paddy, Yunnan, southwest China

QUESTION

At the beginning of this chapter (page 128), we described some features that are special to humans.

Go back to this paragraph and make a list of these specially human features. Then review the successive stages of hominid evolution and identify the particular *Homo* species that first showed these different features. For example – when did hominids start to walk upright and when did they start to develop the beginnings of language? You can also try to make a link to the *evidence* used to support your suggestions.

In making your review of the stages of hominid evolution, you might find it helpful to refer to the list of trends summarised in the bullet points on page 145.

The practice of farming represents a major change in the relationship between early people and their environment. Nomadic hunter-gatherers take from their surroundings and their mobility allows them to move on and exploit fresh areas. Neolithic people, as they settled down, were beginning to interact in a different way with their environment. They manipulated plants and animals for their own benefit and exerted planned control over small areas in their production of food. They cared for their plants and animals – the beginnings of the practice described as **husbandry**; they then harvested their crops and found ways of storing their produce for another season when food may have been less plentiful. They deliberately saved seed to plant for the next season's crop and ensured they kept stock animals for breeding. They are likely to have selected those with the most desirable characteristics – already there was probably a desire for higher yields or different varieties. They were developing the skills of selective breeding – changing the characteristics of the plants and animals they used. The practice of settled farming meant that food had to be stored. Tools were developed to become implements used in the working of the soil or harvesting and processing of the crop. Moving from place to place would have become complicated because of the accumulation of possessions, so that settled community life and agriculture went together in having a profound effect on the future development of people and society.

Finally, we will attempt to put into perspective the changes that have taken place in the evolution of modern humans from our ape-like ancestors. Molecular studies indicate the difference in genome between humans and chimpanzees to be a mere 2 per cent yet the difference in both capability and achievement appears huge. Remember the timescale of human evolution and think first of the enormous rate of change and development of agriculture. The technology now utilised in modern intensive farming has developed in just 10 000 years since the start of the Neolithic culture (about 400 generations ago). Contrast this with the comparatively slow evolutionary progress from humans using the first Oldowan toolkits through Acheulian to Mousterian and finally the microliths in the late Palaeolithic era which spanned over 1 million years (40 000 generations). The biological story of evolution continues and diversifies into the history of human societies, of artistic expression and religions, of scientific discovery and increasing control by humans of the environment. It is appropriate to end this review of human evolution with a quotation from Chris Stringer, a research worker in the field of human evolution:

'How on earth could an animal that struggled for survival like any other creature and whose time was absorbed in a constant searching for meat, nuts and tubers, and who had to maintain constant vigilance against predators, develop the mental hard-wiring needed by a nuclear physicist or astronomer?'

Human populations

World trends in population size

The world population has grown very slowly for most of the 0.5 million years of human existence, but in the last 250 years there has been an overwhelming increase in population numbers. To use rough figures, there were about 1 billion people in 1830, 2 billion in 1930, and 4 billion in 1975. By the turn of the century in 2000, there were just over 6 billion, with a likely 8 billion by 2030. Early figures, long before records were kept, can only be estimates whereas recent figures, based on surveys and censuses, are more reliable. The unmistakable trend is of increased population on a world scale (Figure 11.1). This brings with it serious questions regarding the future with respect to pressure on space and availability of food and the other resources required to support the demands of an expanding population.

QUESTIONS

The first census of a national population took place in Sweden in the 18th century, followed by other countries in Europe during the 19th century. Now, in the 21st century, some less developed countries still lack accurate figures.

- List some of the practical problems to be faced when doing an accurate census of population numbers. Think about nomadic people, homeless people, literacy, suspicion and so on.
- What sort of evidence can be used to estimate world population before records were kept?

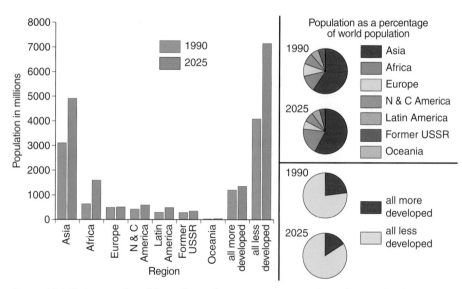

Figure 11.1 Estimates of world population by region in 1990, with predictions for the year 2025.

If we look more closely at the growth of world human population over the last 350 years, we can tease out different patterns in different regions. As a broad generalisation, the more developed countries were the first to show the beginnings of a rapid population increase followed in succession by a number of the less developed countries (Figure 11.2). Growth in the developed countries coincided with the start of the Industrial Revolution. In 19th-century Europe, the Industrial Revolution produced an increase in economic activity with development of technological industries and changes in agricultural practices. At the same time there were fundamental changes in social structures and way of life as a greater proportion of people became wage-earners and thus had less direct dependence on the land and what it could produce. The combined effect of these circumstances undoubtedly contributed to the noticeable increase in

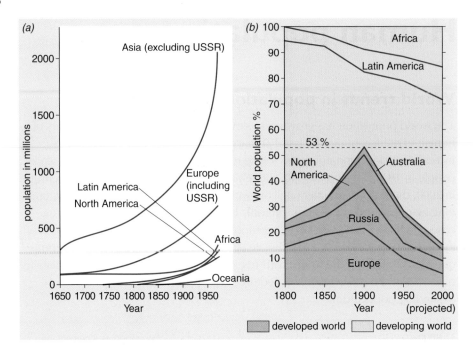

Figure 11.2 Comparing population growth in developed and developing countries:
(a) population growth in major regions of the world, from 1650 to the late 20th century;
(b) share of world population between developed and developing countries, from 1800 to the
late 20th century. Developed countries showed an earlier increase in population growth
(during the 19th century) whereas in developing countries similar increases have occurred in
the 20th century. Diagram (b) shows how the developed world expanded rapidly between 1800
and 1900 but since that time the population in the less-developed world has markedly
increased in numbers and at the same time increased its share of the world population.

QUESTION

Which areas in the world do you expect to have the lowest population densities and why? (You can look back to the chapter on Human ecology in *Exchange and Transport, Energy and Ecosystems*.)

growth rate of the population. A similar, but not identical pattern of change has occurred in less developed countries during the 20th century.

Factors affecting size and growth of human populations

As an introduction to the factors which affect the size and growth of populations, we can look at the relatively simple situation of a population of yeast cells growing in a glucose solution in a flask. Figure 4.2 (Chapter 4) shows how the number of yeast cells increases slowly at first, then rises through the exponential phase to reach a stationary phase followed by a decline and death phase. If we transfer this model to an animal population introduced into a defined area, we would expect a similar initial pattern of growth, with an increase up to a maximum size, which may then fluctuate over a period of time (Figure 11.3). A variety of environmental factors, known collectively as environmental resistance, exerts pressures which limit the size of the population and prevents it continuing to grow indefinitely. Factors that exert such pressure include availability of food, competition, predation and disease. The level of population reached is a measure of the carrying capacity of the area (see Chapter 4). Overall, numbers of this hypothetical animal population are controlled by:

- birth rate – higher birth rate can lead to increase in population numbers
- death rate – lower death rate can lead to increase in population numbers

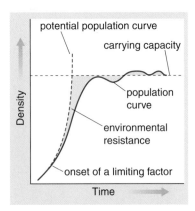

Figure 11.3 Relationship between population growth, environmental resistance and carrying capacity. The carrying capacity is the maximum size of a population that the particular area can support.

- immigration – influx of individuals to area increases population numbers
- emigration – individuals leaving the area decrease the population numbers.

In the case of human populations, social and cultural practices must be included with factors which contribute to 'environmental resistance' and which influence the overall growth of a population.

Even though a prediction of a world population of about 9 billion has been made for 2050, the actual figure may vary considerably and will depend on a whole range of interacting factors. Viewing the possible future trends now, at the turn of the century, it is likely that the greatest growth will continue to come from developing countries, though in some there are indications that the countries are moving towards a more stable population. Factors that are significant include future trends in fertility (hence birth rate) and mortality (death rate), the impact of family planning programmes, changes in the status and education of women and the interaction of poverty and economic growth. Some of these factors are now explored in more detail.

Variations in fertility and birth rates

The term **fertility** can be used generally, in relation to aspects of human reproduction, but more specifically it refers to the actual number of live births achieved in a population over a period of time. The term should be distinguished from **fecundity**, which refers to the physiological ability of a woman (or of a population) to conceive and bear children, otherwise known as the reproductive potential.

Crude birth rate is defined as:

$$\frac{\text{total number of live births in 1 year} \times 1000}{\text{total mid-year population}}$$

Crude birth rate relates the number of births in a year to the whole population, and is expressed by the formula. The crude birth rate includes people who cannot give birth, notably males, and females who are outside the normal reproductive age (generally taken to be between 15 and 49 years). It also includes those unlikely to give birth (because, for example, they are single or infertile). There are considerable variations in the sex and age structure of a population, so other measures of fertility are used which relate the birth rate more specifically to women of child-bearing age or to particular cohorts. Such measures can be useful when comparing demographic trends in different populations.

- **General fertility rate** relates the number of births in a year to the number of women between 15 and 49 years and is expressed by the formula:

$$\frac{\text{number of live births} \times 1000}{\text{number of women aged 15–49 years}}$$

- **Total fertility rate (TFR)** indicates the average number of children that would be born to 1000 women in the current population assuming that each lives to the end of her child-bearing age (15–49 years) and produces children at the same rate as women of these ages did in the year of calculation. It can be interpreted as the average number of live-born children which women would have if they experienced the same age-specific fertility rates of that particular year throughout their lives. A TFR of 2.1 to 2.5 is considered to be the level needed for natural replacement of the population, though this assumes no net migration. Changes in fertility rates in selected countries are shown in Figure 11.5.

- **Gross reproduction rate** is a modification of total fertility rate by giving it in terms of female children only.
- **Net reproduction rate** is a further modification of gross reproduction rate and takes account of women who die before reaching the end of the fertile age range. A net reproduction rate of 1.0 indicates that the population is replacing itself exactly. If less than 1.0, the population is failing to replace itself. If greater than 1.0, the population is likely to increase because the number of potential mothers in the next generation is increasing.

Factors affecting fertility rates

Attempts to explain past patterns of fertility and to predict future trends with any certainty are highly complex because of the interaction of a large number of variables concerning the lives of people. These include social, cultural and religious practices, economic conditions, the level of education and the extent of government controls or intervention. The way that these factors influence fertility will be different in different countries and inevitably change over time. When considering the factors affecting fertility rates, it is appropriate to use the term 'marriage' to include those in stable sexual partnerships as well as formal marriage.

- Proportion of women in marriage:
  Cultural practices relating to birth in or outside marriage vary in different countries and are changing. In Britain, for example, the number of births outside marriage has increased notably in recent years, from 9.8 per cent in 1977 to 40 per cent in 2002.
- Age at marriage:
  In some primitive societies, marriage may occur soon after menarche (first menstruation), giving the potential for many years of child bearing. From the middle of the 20th century, there has been a trend in many countries for later marriage and a later age at birth of first child. Evidence suggests that these are related to there being a higher proportion of educated women, who are perhaps choosing a career rather than marrying and having children early. These women would also need to take a career break to have children, which may contribute to the later age at birth of the first child.
- Number of children within a marriage (family size):
  In traditional societies, high fertility is an advantage, with children providing services such as caring for livestock, weeding crops, carrying wood and water and looking after younger children (Figure 11.4). In recent decades in many developing countries, there has been mass provision of primary schools, so that younger children are no longer available for general agricultural and household chores. Children grow up with different expectations and when they reach adulthood are likely to leave the family home and support an independent family rather than following the traditional way of life. A reduction in family size has been witnessed widely both in developed and developing countries (see Figure 11.5). Figure 11.6 shows changes in family size in England and Wales.
- Control of family size:
  Control or limitation of family size is practised through deliberate measures taken before and after conception, including abstinence, contraception and termination of pregnancy. Success in terms of use of contraceptives is generally linked with the level of development of the country and the extent to which women are educated. In poor countries, cost may be a deterrent. Resistance to the use of contraceptives comes from followers of certain religions, notably Islam and Roman Catholicism. Government family

Figure 11.4 Young children used to care for younger children: two sisters in Sichuan, China

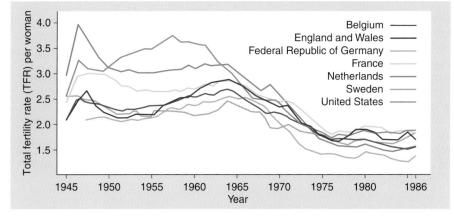

*Figure 11.5 Changes in total fertility
rate in selected countries, from 1945
to 1986. Note the postwar 'baby
boom' in most countries and the
downward trends towards the 1980s.*

planning policies generally encourage limitation of family size, as a means
of controlling growth of the population and so raising the standard of
living. However, a declining population can lead to a diminished labour
force, resulting in reduced economic potential.

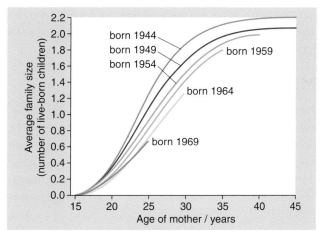

*Figure 11.6 Changes in family size in
England and Wales, for various
cohorts of fertile women born 1944 to
1969. Note the trend to smaller
families as well as the later age at
which mothers give birth and
complete the family – for mothers
born in 1944, at the age of 25 years
the average family size was 1.2 but
2.1 when they were 35, whereas for
mothers born in 1959 the average
sizes were 0.75 and 1.75 respectively.*

EXTENSION MATERIAL

An extreme example of attempts to limit family size (anti-natalist policy) is seen in government policies in China since
1949. Between 1949 and 1990 the population more than doubled, from 540 million to 1134 million
(Figure 11.7). During the 1950s, there was little restraint on family size, the optimistic implication being that the population

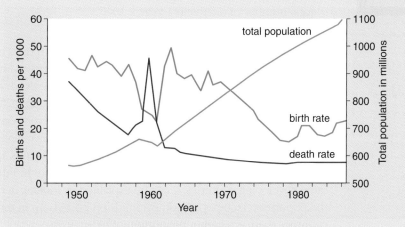

*Figure 11.7 Demographic transition in China, 1949 to 1987, an example of a
government with an anti-natalist policy*

could be fed, however rapid the increase. Widespread famine in the early 1960s led to increased mortality and had a major effect on lowering fertility, though there was a brief 'baby boom' from 1963 to 1965, when food supplies returned to normal. From the late 1960s through the 1970s there was strong government encouragement to reduce family size. Community-based family planning services were expanded and there were campaigns to stress the beneficial effects of a small family in terms of health of mother and children as well as the economic advantages. There were slogans and social pressures from the community to have 'later marriage, longer intervals between births and fewer children'. By 1979 this had been stepped up to the 'one child per family' policy. This introduced incentives for couples to have just one child, but penalties for those having more than two. Success of the policy has been greater in urban than rural communities. Recently, there has been some relaxation of this one-child policy for minority nationalities. This provides a means of positive encouragement for the survival of these groups, some of which have dwindled to very low numbers or been absorbed into the dominant Han Chinese population and so lost their identity.

Effects of infertility

In some couples, one or other partner might be infertile for physiological or genetic reasons. Medical technology has opened up ways to help childless couples overcome infertility through, for example, drugs to stimulate ovulation, *in vitro* fertilisation and artificial insemination. While important to the individuals affected, infertility and its treatment does not have a significant effect on fertility trends in the population as a whole.

Mortality and death rates

Death rates (mortality rates) vary within a population in relation to age, sex, environmental conditions, social factors and disease. Death rates also vary between different countries. Changes in death rates influence the growth of a population. However, when making comparisons of the situation with another, it must be realised that some of the data are unreliable. For example, earlier records in England and Wales were taken from parish registrations of baptisms and probably discounted infant deaths between birth and baptism. Similarly, early deaths of unwanted children are sometimes obscured and even now registration of deaths may be incomplete in some countries.

Crude death rate is defined as the number of deaths per 1000 in the population in a given year, usually taken as the mid-year population and found using this formula:

$$\frac{\text{total number of deaths in 1 year} \times 1000}{\text{total mid-year population}}$$

However, as with crude birth rates, these 'crude' death rates do not allow for differences in composition of the population, particularly with respect to age and sex, both of which affect mortality probabilities. Crude death rates for a number of countries are given in Table 11.1. Generally, crude death rates are higher in developing than in developed countries. It may, however, seem anomalous that certain developed countries (for example, Sweden) have higher death rates than some developing countries (for example, Brazil). Whilst this appears to conflict with the data relating to life expectancy, which is higher in Europe than in these same developing countries, it can be explained by looking at the age structure of the population (see population pyramids on pp 151–2). In the developed countries there is a higher proportion of older people, which accounts for the higher overall death rate. In most countries, death rate of males is higher than that of females, though the biological basis for this difference is not understood.

> ### DEFINITION
>
> **Life expectancy** is the average number of years a person can expect to live, usually expressed at birth, but can be related to any age-cohort.

Table 11.1 *Estimates of death rates, infant mortality rate and expectation of life in selected countries around 1990*

Country	Death rate per 1000 per year	Infant mortality rate per 1000 live births	Expectation of life at birth / years	
			Male	Female
Africa				
Central African Republic	17	100	44	47
Egypt	10	61	59	61
Mali	20	164	46	50
Zimbabwe	10	61	57	60
Europe				
Poland	10	16	67	76
Portugal	10	13	68	75
Sweden	12	6	74	80
UK	12	9	72	78
Asia				
Bangladesh	15	114	57	56
China	7	30	68	71
India	11	94	52	52
Japan	7	4	75	81
Sri Lanka	6	26	68	72

QUESTION

Suggest reasons for the trend in all countries towards lowering of infant mortality rates.

Causes of mortality – patterns and trends

Patterns in the causes of mortality have changed over time and, in the 21st century, continue to show variations in different regions of the world, depending on local conditions or sometimes on isolated events. When reviewing data, it must be appreciated that reporting of deaths may be incomplete and that diagnosis of illnesses may be vague or inaccurate, particularly where medical knowledge is inadequate compared with the best known in the 21st century. In England and Wales, death certificates were first issued in 1850 and this led to greater reliability in the data, but even today in some countries, there is no complete system for recording deaths.

Changes in the causes of death in England and Wales over the 140-year period from 1851 to 1990 are given in Figure 11.8. These data show that deaths from infectious and parasitic diseases and from tuberculosis have undergone a marked decline since 1851 whereas there is a substantial increase in deaths from heart disease, strokes and cancer. When you study demographic trends (pages 160 to 162), you can give further consideration to these changes and link them with the stages in demographic transition represented in Figure 11.11.

QUESTIONS

A cholera epidemic in London in the mid-1850s was linked to a water pump in Soho.

- What steps are taken now to ensure clean water supplies? List other hygiene measures which are likely to have led to a reduction in infectious diseases.
- Find out when the first smallpox vaccinations were carried out and the date when this disease was eradicated on a world-wide scale.
- What methods are used today to control malaria and why are some preventive measures becoming less successful?

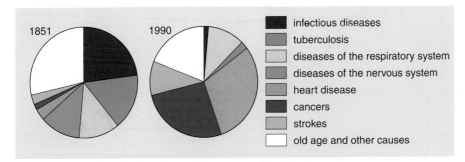

Figure 11.8 Causes of deaths in England and Wales, comparing 1851 with 1990

HUMAN POPULATIONS

Figure 11.9 Public health advice in rural Bhutan – to keep water supplies clean

In looking for explanations for the decline in infectious diseases, we can consider a number of factors.

- **Biological** – such as a reduction in virulence of a microorganism or an increase in natural resistance in the population. These are probably not the main reason for a decline in infectious diseases as they are unlikely to have happened at the same time in a range of microorganisms.
- **Environmental and social** – improvements in living conditions (especially in housing, clean water supplies and nutrition), and better understanding of hygiene (including food preparation and removal of refuse and sewage) leading to less exposure to infectious pathogens. These factors are best summarised as 'improved standard of living'.
- **Medical intervention** – including vaccination and immunisation programmes encouraging prevention, as well as understanding of diseases and medication or treatment which improves chances of curing them.
- **Public health advice and legislation** – to ensure appropriate education of the people and implementation of measures to control or reduce disease (Figure 11.9).

In the late 20th century, heart disease and cancers were the main causes of death in developed countries. This represents a shift towards chronic or deteriorative diseases, associated with older adulthood and a more affluent lifestyle. To put it in simple terms, this can be seen as a consequence of there being fewer deaths from infectious diseases. We should, however, remember that many of the less developed countries are situated in warm latitudes where the climate encourages spread of diseases such as yellow fever, malaria and bilharzia. In these countries, improvements in housing, sanitation and general standard of living, together with public health and improved medical facilities, are likely to bring about a similar reduction in mortality.

Population growth curves and demographic trends

Different attempts to describe and interpret patterns of population growth have led to a number of models. The demographic transition model (Figure 11.10), has been used to represent changes in recent European populations. This model identifies four main stages, related to birth rate and death rate. The term **crude birth rate** refers to the number of births per 1000 persons in a year and **crude death rate** is expressed as the number of deaths per 1000 persons per year. In this chapter, these are referred to as 'birth rate' and 'death rate' respectively. The four main stages are:

QUESTION

How far do you think events occurring at the time of the Industrial Revolution became a means of increasing the carrying capacity of the land (see page 154 and Figure 11.3)?

Figure 11.10 The demographic transition model of population growth. The terms 'high stationary phase' and 'low stationary phase' refer to the birth and death rates rather than to the numbers in the population.

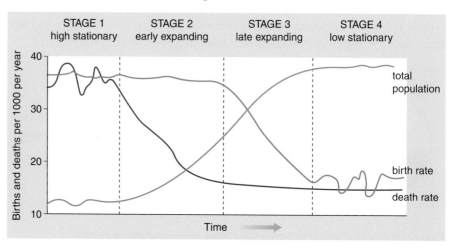

160

- **Stage 1** – births approximately equal deaths in a year. Birth and death rates are both high and the population remains steady at a low level (with some fluctuations).
- **Stage 2** – shows the start of a rapidly expanding population, linked mainly to a decline in death rate.
- **Stage 3** – shows a decline in birth rate associated with an increasingly urban society in which the economic value of a large family becomes less important.
- **Stage 4** – has low birth rate and low death rate, giving a second relatively stable phase.

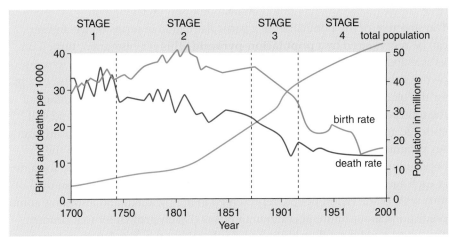

Figure 11.11 Demographic evolution in England and Wales since 1700. Stages 1, 2, 3 and 4 of the demographic transition model are recognisable.

The demographic transition model should be treated as a generalisation, which does not necessarily apply to all nations but gives a useful overall summary of trends in population growth. A similar pattern is recognised in some developing countries in the 20th century.

Demographic changes in England and Wales since 1700 are summarised in Figure 11.11 where stages of the demographic transition model can be recognised. Stage 1 ended around the 1740s and showed typical high and fluctuating birth and death rates with a long-term slow increase in the population. Stage 2 lasted from the 1740s through to the 1880s. Research has suggested that this population increase is accounted for mainly by changes in fertility linked to the timing of marriage. When economic conditions were favourable and wages rose, there were more marriages, people tended to marry at a younger age and to produce larger families, and hence to have more births per family.

Stage 3 is characterised by a fall in both death and birth rates. The latter may be linked to continued improvement in living standards together with improved methods of birth control, which gave parents greater freedom to restrict the size of their families if they wished. The noticeable drop in death rate in the late 19th century was due to a decrease in infectious diseases, such as typhus, smallpox and diphtheria. Because of improved nutrition and increased prosperity, people showed greater resistance to disease. Better hygiene and the provision of urban water supplies and drainage systems contributed to the reduction of disease. Towards the end of Stage 3, the effect of the First World War on both death and birth rates can be seen. Since the 1920s, low but fluctuating birth and death rates have led to a slowing down of population growth. A low birth rate during the 'depression' years in the 1930s is contrasted with a postwar 'bulge'.

The natural increase of the population in an area is given by the excess of births over deaths: the wider the gap, the greater the increase. The natural increase contributes to the growth rate of the country, but migration of people into and out of a country (immigration and emigration respectively) must also be taken into consideration. For world population, migration is not relevant. Migration is discussed in more detail on page 163.

Population structure – population pyramids

The structure of the population of a particular nation in terms of age and sex can be represented by a **population pyramid**. This is a histogram, turned on its side and placed back to back, with males on the left-hand side and females on the right. The population is generally presented in 5-year age groups, except for the oldest (say 85+ years), and each group is expressed as a percentage of the total population. New births are added on at the base of the pyramid and each age group moves up as time passes. Population pyramids, also called 'age-sex pyramids', give a profile of the distribution of different ages within the population (in that particular year) and the balance between males and females in each age group. The pyramids reflect history of recent events (within the time scale of the oldest people represented in the pyramid) and also allow some predictions about future population trends and likely growth for that country. The actual numbers reflect births and deaths within the population and are also affected by migration into and out of the area represented by the pyramid.

Population pyramids for England and Wales in 1881, 1931 and 1986 are shown in Figure 11.12. These illustrate different situations with respect to the structure of the population and can be related to the stages of demographic transition occurring over these years (see Figure 11.11). The 1881 pyramid is typical of a population with a high death rate or at an early stage of demographic evolution. It shows a young population, with 70 per cent under the age of 35. The 1931 pyramid shows a population with a reduced birth rate and reflects an older population with increased life expectancy. In the 1986 pyramid, the narrow base shows a continuing reduction in birth rate and the higher numbers over 65 years (particularly of females) indicate an increasing life expectancy. In the 1986 pyramid, there are two bulges, one in the 35- to 40-year age range and also in the 20- to 25-year age range. The first group was born in the late 1940s to early 1950s, in what is known as the postwar 'baby boom'. The second group was born when the earlier 'baby boom' group reached reproductive age.

Figure 11.12 Population pyramids for England and Wales in 1881, 1931 and 1986. The data are arranged in 5-year age groups (vertical axis) and represented as a percentage of the total population (horizontal axis). The oldest age group is an exception as it contains all over the age of 85. Other population pyramids in this chapter are represented in a similar way.

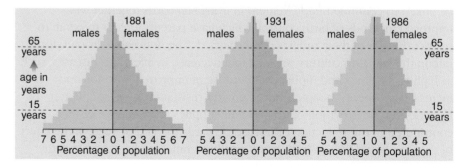

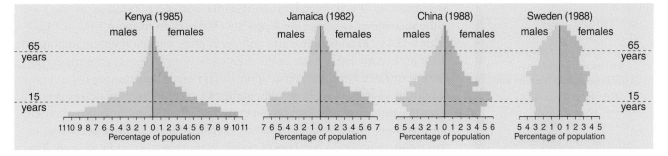

To understand how to interpret population pyramids, we can look at some specific examples in a range of countries (see Figures 11.13). This should help you see the sort of information that can be obtained from these pyramids and illustrates how populations can be described as **stable**, **increasing** and **declining**.

The markedly triangular shape of the pyramid for Kenya (Figure 11.13) is typical of a developing country. The broad base indicates a high birth rate and only a relatively small proportion of the people survive into old age (taken as over 65 years). While the high birth rate gives the potential for a rapidly **increasing population**, if death (mortality) rates are high, particularly in the early years, numbers are kept in check and the triangle tapers towards a point. If, however, say through improved medical care, there are lower death rates in early years, there is likely to be less marked tapering of the shape of the pyramid and this situation would give rise to an increase in population numbers. Pyramids that are triangular but with a narrower base than that shown by Kenya (e.g. China) suggest a reduction in births (due, perhaps, to implementation of family planning policies) and a reduction in child mortality (probably through improved medical care).

Pyramids that have more or less parallel sides are typical of more developed countries. These show a relatively **stable population**, with similar birth rates over successive years and without a noticeable loss of numbers through deaths in the early years. A higher proportion of the population survives into old age (over 65 years). However, if the base becomes narrower, this could result in a situation in which there may be too few births to replace the population, leading to a **declining population**.

An uncharacteristic bulge or hollow in the vertical sides of the pyramid reflects an earlier sudden increase in births over a limited number of years (such as a post-war 'baby boom'), or a sudden high death rate (say through war or starvation), again over a limited number of years.

Migration

Migration within a nation state (for example from rural to urban communities, or nomadic peoples moving between lowlands and highlands) has no effect on the overall population statistics of a nation and migration has no relevance to global population. However, migration across international boundaries can have an effect on the population growth curves and on the population pyramids of a nation (see page 162).

Figure 11.13 Population pyramids for Kenya, Jamaica, China and Sweden. See caption for Figure 11.11 for explanation, but note that the oldest age group for the Kenya pyramid includes all over the age of 80.

QUESTIONS

Look at the population pyramids in Figure 11.12 and answer these questions about populations.

- Which countries are at an early stage of demographic transition? Are these 'developed' or 'developing' countries?

- In 1985, what percentage of the population of Kenya was under the age of 35?

- What deductions can you make about birth rate in relation to the economic status or social systems of the country? Suggest some reasons for these relationships.

- What differences can you see in the balance between males and females in different age groups – around birth, up to 15 years, 15 to 49, over 65?

HUMAN POPULATIONS

QUESTIONS

- List reasons for changes in fertility and population increase during the 19th and 20th centuries.
- What are some of the pressures on the *environment* (at a local and at an international level) arising from population growth?
- Suggest reasons why *governments* may wish to minimise or to increase the nation's population growth.
- Think about why *individuals* may wish to limit or control their family size.

As an example of how migration can affect the size of a population, we can look at the situation in Jamaica, over a period of 90 years, from about 1892 to 1982 (Figure 11.13). In addition to birth and death rates, migration was an important factor in determining the size of the population. Employment problems in the 1950s stimulated large-scale emigration, the people being attracted by opportunities elsewhere. Eventually, immigration controls in the destination countries had a limiting effect on the emigration from Jamaica. Those who did emigrate were mainly young adults, so, 30 years later (in 1982), there was a noticeable reduction in numbers of people in the age group over 45 years. This emigration also meant that there were fewer people of reproductive age, resulting in a reduced birth rate, hence a lowering in the growth rate of the population.

EXTENSION MATERIAL

Migration of people over a longer timescale

There have been repeated examples of movements of people from one area to another. Such migrations started perhaps 100 000 years ago with the earliest populations of *Homo sapiens* (see Chapter 10). Possible disperal routes taken by our ancestors as they spread from their origin in Africa to all continents of the globe are shown in Figure 11.14. Nomadic hunter-gatherers certainly would have moved short distances as they exploited the resources of their surrounding environment. Indeed, our estimates of early population numbers are based on the carrying capacity of the land, or the maximum population density that could be sustained by the way of life of the people in the area. As resources diminished, it is likely that the people moved elsewhere. Transition to settled agriculture in the Neolithic Age inevitably increased the potential productivity of the land for the people and was probably associated with a surge in population numbers. At that time, there were no national or international boundaries, but as nation states became identified, we can trace large-scale migrations of people which have had noticeable effects on the population numbers of a particular area or country.

The impetus to migrate is usually linked to pressure at home, often triggered by increasing population, crop failure and scarcity of food. At the start of the 21st century we continue to see the political pressures linked to migrants seeking refuge (often as asylum seekers) in countries distant from their own.

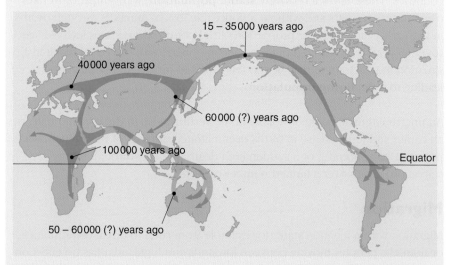

Figure 11.14 Possible dispersal routes of Homo sapiens, *from Africa around the globe over the last 100 000 years. The map has been reconstructed using evidence from fossils and from genes.*

Implications of world population trends

Graphs of population numbers suggest that the human population is now in the exponential phase of growth (Figure 4.2) and, in the early 21st century, there is little sign that we are approaching the stationary, let alone the declining phase. The growth rate in some developed countries is relatively low but it remains higher in many developing countries. Even if extreme measures (such as China's 'one-child' policy) were taken on a global scale in an attempt to limit population growth, with the numbers present in the existing population, continuation at a very low growth rate would still add huge numbers to the world population over the next few decades. Predictions that suggest the world population will reach 8 billion in 2030 assume that 'environmental resistance' will not exert sufficient effect to limit its growth.

Availability of food is one of the factors that controls or limits the growth of a population, leading towards the stationary or decline phase when supplies of nutrients become low. This applies to a yeast population growing in glucose in a flask, or indeed to any animal population in a given area. In humans, the population size of early hunter-gatherers must have been limited by the area within which they could hunt for food and on the availability of natural food sources in that area. The transition of these early human hunter-gatherers to settled agriculture in the Neolithic Age was almost certainly associated with a surge in population numbers. The practice of growing crops, harvesting and storing excess through a lean season, together with provision of a reliable source of meat and other products from domesticated animals, offered enormous potential for increasing the food available in a given area.

Over the centuries, developments in agriculture have been a means of increasing the carrying capacity of the land with respect to human demands. Selection of favourable characteristics in crops has resulted in improved varieties with higher yields and tolerance of a wide range of environmental conditions. Similarly, breeds of domesticated animals have been developed and provide humans with a rich range of meat (and eggs and milk) in far greater abundance than our Neolithic ancestors could have imagined. Gradual introduction of mechanisation, now seen in the form of tractors, machinery for irrigation and every stage of crop cultivation through to harvesters, has given a further boost to the potential yield that farmers can get from the land.

QUESTIONS

Questions about food production

- List some ways that people can *increase* the carrying capacity of the land for human populations. Think about practices in modern agriculture (scientific and technological applications), biotechnology (old and new) and possible future developments.
- Then list some human activities that effectively *decrease* the potential of land to produce food.
- What sacrifices are made in the drive for increased food production? Some points to consider include animal welfare, quality, taste, diminished gene pool, loss of natural habitats, use of energy.

HUMAN POPULATIONS

Figure 11.15 Intensive glasshouse culture: tomatoes grown by hydroponics (Kent, UK)

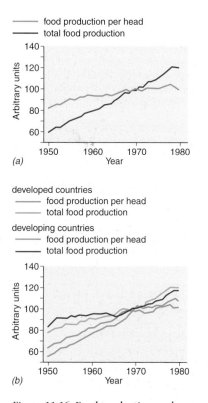

Figure 11.16 Food production and the increasing population: (a) total world production of food more than doubled between 1950 and 1980; (b) comparison of developed and developing countries indicates that the developing countries showed a larger percentage increase than developed countries. The year 1970/71 is given an arbitrary value of 100.

At the beginning of the 21st century, pressure on land is becoming acute. Whilst there is still some scope for bringing fresh land into cultivation, parts of the land surface across the globe are unsuitable for cultivation because of the topography, the nature of the soil or the climate. In some regions, particularly Europe and Asia, a very high proportion of potential arable land is already under cultivation. Social factors may discourage people from cultivating land in remote areas. As populations expand, often the best agricultural land is swallowed up in housing and the supporting infrastructure of urbanisation.

Intensive farming methods strive to maximise productivity and increase the yield from agricultural land. Artificial fertilisers are used to increase the nutrient content of the soil (or replace that which was lost when crops were removed); pesticides help to reduce crop losses from pests and disease. Control of the environment by cultivation in glasshouses is a way of extending the growing season and allows the harvest to be more predictable in terms of quality, quantity and timing (Figure 11.15). In rearing animals for meat, production has increased through control of diets, manipulation of reproduction and dense stocking in housing where environmental conditions can be kept at their optimum. There is further potential for increasing production by culture of microorganisms. Protein or other organic molecules derived from these sources can be introduced into novel foods to become part of the human diet. Application of gene technology gives scope for being very specific about the precise characteristics to be introduced into the organism being produced for food. Modern methods of packaging and storage of foods have contributed to a considerable reduction in the wastage of food after harvest through deterioration and spoilage.

Despite the success of applying science and technology, there is a global imbalance in the availability and distribution of food and there is concern whether production can continue to keep pace with the increasing number of mouths to feed in the 21st century. There is little doubt that in affluent, well-developed societies, there is an abundance of high-quality food whilst at the same time, in other places, there are people living on minimal diets, suffering from hunger and malnutrition with some close to starvation (Figure 11.16). This may be the result of crop failure in subsistence areas, or perhaps due to the effect of wars, or simply a consequence of poverty within an otherwise prosperous society. As income per head increases in developing countries, there is a corresponding demand for more food, in terms of quality, quantity and variety. This puts further pressure on the capacity of the land to fulfil these demands.

Some of the consequences of human activities on the environment in the context of increasing world population are discussed in Chapter 5 and also in *Exchange and Transport, Energy and Ecosystems* Chapters 8 and 9. Consideration is given to changes in land use, the impact on natural ecosystems and limitations of natural resources, including sources of energy and increasing demand for energy use. There are also effects of pollution as a result of human activity. In terms of feeding the world population, the emphasis must be to find ways of increasing food production in a sustainable way that maintain long-term ecological stability. 'Progress' and changes are inevitable. However, the ultimate resource is people and we must be optimistic that, as in the past, human ingenuity will find ways that will continue to provide for and support the increasing demands of the human population in the future.

QUESTION

As well as there being more mouths to feed in the future, the composition of populations is also changing, particularly with respect to the age structure.
- What are the consequences of an increasingly older population in developed societies?
- As medical facilities become more widely used in less developed countries, what effects will this have on the age structure – now, in 10 years time, and in 50 years time? What further demands will these countries make on world resources, such as food and energy?

ADDITIONAL MATERIAL

Assisted reproduction – artificial insemination and *in vitro* fertilisation

Louise Brown, born in 1978, was the first 'test-tube' baby. Conception was assisted by *in vitro fertilisation (IVF)* and her birth represented a landmark in the development of techniques which can help infertile couples. **Artificial insemination (AI)** is used routinely in animal husbandry, particularly in cattle breeding programmes: its application to humans has made important contributions to successful conception in otherwise infertile couples. The steps in the procedure for IVF are summarised in Figure 11.17.

Within the framework of IVF, there are variations in how and when assistance is given, depending on the cause of infertility. Some of these are outlined below:

- **gamete intrafallopian transfer (GIFT)** – the ova and sperm are both collected and then placed together in the oviduct (Fallopian tube), where fertilisation may take place, followed by implantation in the endometrium
- **zygote intrafallopian transfer (ZIFT)** – fertilisation occurs externally, but the zygote is then immediately placed in the oviduct where cell division starts, followed by movement to the uterus for implantation
- **oocyte donation** – when an 'infertile' woman

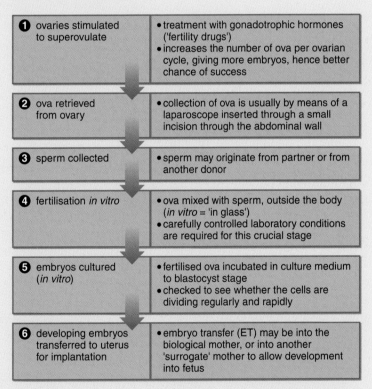

Figure 11.17 Outline of steps in procedure for in vitro *fertilisation (IVF)*

cannot produce ova (oocytes), donated ova from another woman may be fertilised by the partner's sperm. The embryo can then be placed into the uterus of the infertile woman, where development and pregnancy may continue normally. The child would be biologically related to one of the parents.

Artificial insemination may be carried out with the partner's (husband's) sperm (**HI**) or with sperm derived from a donor (**DI**). In the latter case, the child would be biologically related to one of the parents, and this may be an acceptable course of action where the man has a very low sperm count. When collecting semen from a man for artificial insemination, the semen is generally centrifuged and treated so that the concentration of sperm is increased and the sample checked to ensure it contains adequate numbers of high quality sperm with suitable motility. It is introduced by a small syringe, either into the woman's vagina or higher through the cervix into the uterus. AI may be successful when the woman's cervix has a hostile reaction to sperm, if the man has a low sperm count, or when he is unable to have normal intercourse, for example, because of spinal injury.

Success rates in assisted reproduction are still relatively low: GIFT ~18 to 26 per cent, ZIFT ~17 per cent and IVF ~13 per cent per cycle (some clinics claim up to 30 per cent, but this may be linked to careful selection of potential parents). While successful conception may bring happiness to otherwise infertile couples, the practices of assisted reproduction and research with human embryos represent an area of extreme sensitivity which is ethically controversial. Some people, for religious or personal reasons, disagree with any form of contraception, abortion or interference with the so-called natural processes of reproduction. There is the danger that, if used irresponsibly, there could be selection of sexes, which would upset the balance within a population, or indiscriminate use of donor gametes could accidentally result in biologically close relatives producing children. Traditional taboos against incest or marriage between close relatives are an ancient way of minimising the chance of genetic defects appearing within a family. In the UK, the Human Fertilisation and Embryology Act in 1990 laid down guidelines for legal aspects of children produced by artificial insemination and IVF and also established the Human Fertilisation Embryology Authority (HFEA). The legal parents of children produced by assisted conception are not necessarily the biological parents and at some stage, the children may wish to know their true biological parents. The HFEA requires registration of all people whose gametes are used for assisted reproduction and all children produced by assisted reproduction and also controls and reviews research with human embryos.

Assessment questions

The following questions have been chosen from recent Unit tests on the content of Units 5B and 5H of the Edexcel Biology and Biology (Human) Advanced GCE specification. The style and format of these questions will be similar to those in future tests. Shorter questions are designed to test mainly knowledge and understanding of the topics and the longer questions contain sections in which you may be required to demonstrate skills of interpretation and the evaluation of data. Some sections of the longer questions may require extended answers, for 4 or more marks.

The written tests of these Units have synoptic sections, containing questions designed to give students the opportunity to make connections between at least two units of the specification, and to use skills and ideas developed throughout the course in new contexts. Some synoptic questions have been included together with questions from Synoptic papers relevant to the topics in Units 5B and 5H.

As it is difficult to include questions that cover all the topics in these Units, extra practice questions may be found on past papers for the examination that preceded Curriculum 2000. You should check that the content of these is relevant to the present specification.

Chapter 1

1   Photosynthesis is a complex metabolic process which can be influenced by many different environmental factors.

   (*a*) Explain the term **limiting factor** with reference to photosynthesis. **[2]**

   (*b*) An investigation into the effect of light intensity and carbon dioxide concentration on photosynthesis was carried out using pond weed. The pond weed was placed in a test tube that contained pond water and a quantity of sodium hydrogencarbonate. The light was provided by a lamp. The oxygen bubbles produced by the pond weed were directed into a length of capillary tubing.

   The graph below shows how the rate of oxygen production of the pond weed changed with light intensity when immersed in two different concentrations of sodium hydrogencarbonate.

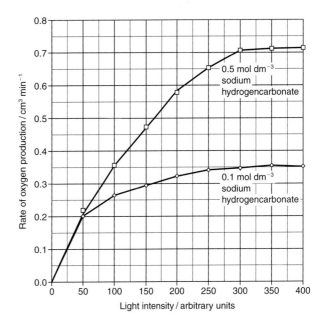

   (i)   Explain the shape of the graph between a light intensity of 0 and 400 arbitrary units in the sodium hydrogencarbonate concentration of 0.5 mol dm^{-3}. **[4]**

   (ii) Describe and explain the effect of increasing the concentration of sodium hydrogencarbonate on the rate of oxygen production. **[3]**

   (iii) A number of precautions would need to be taken while carrying out this investigation in order to obtain reliable data. Describe **one** precaution and explain why this precaution is necessary. **[2]**

   (*c*) During the light-dependent stage of photosynthesis, oxygen is produced by photolysis. Describe the process of photolysis and explain its role. **[3]**

   (Total 14 marks)

   (Edexcel GCE Biology Unit Test 6105/01, June 2002)

ASSESSMENT QUESTIONS

2 (*a*) The rate of photosynthesis can be limited by a number of factors.

Explain why temperature can be a limiting factor in photosynthesis. **[2]**

(*b*) The diagram below shows the structure of a chloroplast.

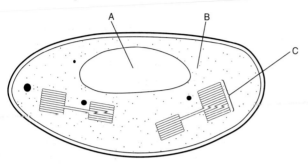

Name the parts labelled A, B and C. **[2]**

(*c*) The flow diagram below shows some of the processes which occur in the light-independent reaction of photosynthesis.

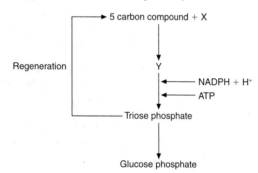

(i) Name the substances represented by the letters X and Y. **[2]**

(ii) State the origin of the NADPH + H$^+$ and the ATP used in the light-independent reaction. **[1]**

(*d*) Describe how the products of photosynthesis are transported in the plant. **[5]**

(Total 12 marks)

(Edexcel GCE Biology Unit Test 6105/01, January 2003)

3 The graph below shows the absorption spectra for three photosynthetic pigments found in a leaf.

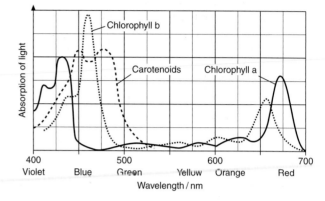

(*a*) Describe the absorption spectrum of chlorophyll a. **[2]**

(*b*) Using the information in the graph, explain why it is advantageous for plants to have more than one type of photosynthetic pigment. **[2]**

(*c*) Describe a technique that you would use to separate the photosynthetic pigments found in chloroplasts. **[4]**

(Total 8 marks)

(Edexcel GCE Biology Unit Test 6105/01, June 2003)

Chapter 2

1 The table below refers to growth substances and their functions or effects.

Complete the table by inserting suitable words in the blank spaces.

Plant growth substance	One function or effect
Auxin	
	Promotes cell division
Gibberellin	
	Causes ripening in fruits such as bananas and tomatoes
Abscisic acid (ABA)	

(Total 5 marks)

(Edexcel GCE Biology Unit Test 6105/01, January 2003)

Chapters 3, 4 and 5

1 The chimpanzee *Pan troglodytes* is classified as shown in the table below. Complete the table by inserting the appropriate word in each of the spaces.

Kingdom	Animalia
	Chordata
Class	Mammalia
	Primates
Family	Pongidae
Genus	
	troglodytes

(Total 4 marks)

(Edexcel GCE Biology Unit Test 6105/01, June 2002)

2 (*a*) Explain the term **net primary production**. **[2]**

(*b*) An area of deciduous forest in North America was destroyed by fire in order to create new farmland. Within a few years the land was abandoned. The graph below shows the change in the net primary production of the land and plant biomass over a period of 160 years, from the time the farmland was abandoned.

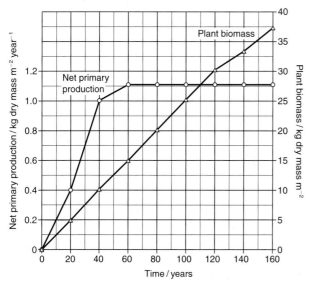

(i) Once the land was abandoned, succession took place. Describe the changes that you would have expected to occur over the 160 years. **[3]**

(ii) Calculate the percentage increase in biomass between 40 and 100 years. Show your working. **[1]**

(iii) The units of plant biomass are kg dry mass m^{-2}. Explain why it would be more informative to have determined the energy content of the plants. **[2]**

(iv) Suggest why the biomass continued to increase over the 160 year period but the net primary production levelled off after 40 years. **[3]**

(Total 11 marks)

(Edexcel GCE Biology Unit Test 6105/01, June 2002)

3 During the 1940s and 1950s Clear Lake in California was plagued with swarms of midges (small flies). In 1949, the lake was sprayed with a pesticide that contained DDT in order to kill all the midges. When the lake was sprayed again in 1954 and 1957 many grebes (a type of fish-eating bird) died. Analysis revealed that the fish in the lake were heavily contaminated with the pesticide.

The flow chart below shows the build up of pesticide in the organisms living in the lake in the late 1950s.

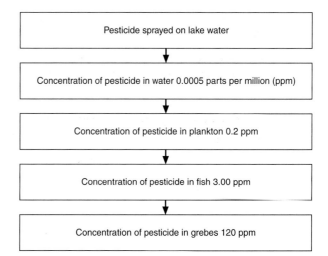

(a) DDT is a non-biodegradable pesticide. Explain the meaning of the term **non-biodegradable**. [2]

(b) Explain why the pesticide concentration in the plankton was only 0.2 ppm while the concentration in the grebes was 120 ppm. [3]

(c) Suggest and explain how the midges could have been controlled without using pesticides. [2]

(Total 7 marks)

(Edexcel GCE Biology Unit Test 6105/01, January 2003)

4 Whiteflies (*Trialeurodes vaporariorum*) are a pest of a number of different glasshouse crops, such as tomatoes. The number of whiteflies can be controlled by introducing the parasitic wasp *Encarsia formosa* into the glasshouse. This is an example of the biological control of insect pests.

(a) Explain what is meant by the following terms.

(i) Biological control [2]

(ii) Parasitism [2]

(b) The rate of reproduction in *Encarsia formosa* is influenced by temperature, as shown in the table below.

Temperature / °C	Rate of reproduction of wasps
below 10	Much lower than that of whitefly
18	Equal to that of whitefly
27	Much higher than that of whitefly

Suggest why it could be ineffective to use this wasp as a form of biological control in each of the following.

(i) A glasshouse maintained at a temperature below 10 °C [2]

(ii) A glasshouse maintained at a constant temperature of 27 °C [2]

(Total 8 marks)

(Edexcel GCE Biology and Biology (Human) Unit Test 6106/03, June 2002)

Chapter 6

1 (a) With reference to examples, explain what is meant by each of the following terms.

(i) Continuous variation

(ii) Discontinuous variation [4]

(b) Variation can arise by point mutations.

(i) Describe **two** types of point mutation which could result in the changing of a GCT codon to a GCA codon [2]

(ii) Suggest why such a point mutation might have no effect on the phenotype. [2]

(c) Variation occurs in humans in their ability to detect sound. Two unlinked genes, each with two alleles (**A** and **a**, **B** and **b**) affect hearing in humans. A person who is homozygous recessive for either or both of these genes is deaf.

A couple have genotypes **Aabb** and **aaBb**. Using a genetic diagram, determine the probability that a child produced by them will have **normal** hearing. [4]

(Total 12 marks)

(Edexcel GCE Biology Unit Test 6105/01, January 2003)

2 (a) Explain what is meant by the term **multiple alleles**. [2]

(b) The diagram below shows a family tree in which the ABO blood-group phenotypes are shown for some members of the family.

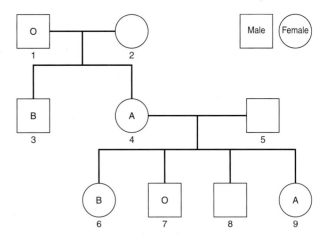

(i) Using the symbols I^A, I^B and I^O to represent the alleles, indicate the genotypes of the following family members: Member 2 and Member 5. [1]

(ii) Using a genetic diagram, show the possible blood groups for family Member 8. **[3]**

(Total 6 marks)

(Edexcel GCE Biology Unit Test 6105/01, June 2002)

Chapter 7

1 Write an essay on fertilisation in humans and the detection of fetal abnormalities.

(Total 15 marks)

(Edexcel GCE Biology and Biology (Human) Unit Test 6106/03, June 2003)

Chapters 8 and 9

1 Using gene technology techniques, a gene that confers herbicide resistance can be incorporated into the DNA of economically important plants such as wheat and tobacco. The flow diagram below outlines such a technique.

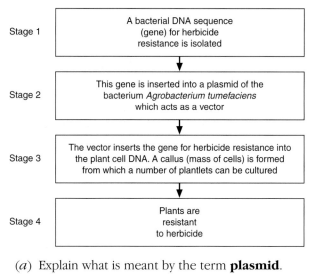

Stage 1	A bacterial DNA sequence (gene) for herbicide resistance is isolated
Stage 2	This gene is inserted into a plasmid of the bacterium *Agrobacterium tumefaciens* which acts as a vector
Stage 3	The vector inserts the gene for herbicide resistance into the plant cell DNA. A callus (mass of cells) is formed from which a number of plantlets can be cultured
Stage 4	Plants are resistant to herbicide

(*a*) Explain what is meant by the term **plasmid**. **[3]**

(*b*) A marker gene may be inserted into the plasmid together with the herbicide gene. Explain the reason for this. **[2]**

(*c*) The gene for herbicide resistance controls the synthesis of a new protein. Describe how the information in this gene is transferred to mRNA during protein synthesis. **[4]**

(*d*) Suggest possible disadvantages that this example of gene technology may have. **[3]**

(Total 12 marks)

(Edexcel GCE Biology Unit Test 6105/01, June 2002)

2 (*a*) Explain what is meant by a **genetically modified organism**. **[3]**

(*b*) The flow diagram below shows some of the stages in the production of the enzyme chymosin using a genetically modified microorganism. This enzyme is first synthesised as an inactive precursor, prochymosin, which is then converted to chymosin.

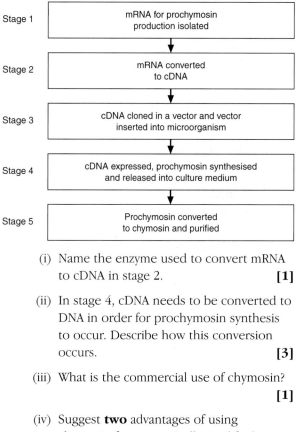

Stage 1	mRNA for prochymosin production isolated
Stage 2	mRNA converted to cDNA
Stage 3	cDNA cloned in a vector and vector inserted into microorganism
Stage 4	cDNA expressed, prochymosin synthesised and released into culture medium
Stage 5	Prochymosin converted to chymosin and purified

(i) Name the enzyme used to convert mRNA to cDNA in stage 2. **[1]**

(ii) In stage 4, cDNA needs to be converted to DNA in order for prochymosin synthesis to occur. Describe how this conversion occurs. **[3]**

(iii) What is the commercial use of chymosin? **[1]**

(iv) Suggest **two** advantages of using chymosin from genetically modified organisms in commercial processes, rather than mammalian rennin. **[2]**

(Total 10 marks)

(Edexcel GCE Biology Unit Test 6105/01, June 2003)

ASSESSMENT QUESTIONS

Chapters 10 and 11

1 Complete the table below by stating **two** external features that are characteristic of each group.

Group	External feature 1	External feature 2
Apes		
New world monkeys		

(Total 4 marks)

(Edexcel GCE Biology (Human) Unit Test 6115/01, June 2002)

2 Mitochondria contain DNA which is useful for evolution studies. Within this DNA, point mutations are common. These mutations can be used as a 'molecular clock' to measure human evolution. Scientists have estimated that one point mutation in the mitochondrial DNA represents approximately 21 500 years.

When two groups split off from a common ancestor, each accumulates different mutations from that time on. The number of mutations is proportional to the length of time the groups have been separate.

(*a*) Explain what is meant by the term **point mutation**. **[2]**

(*b*) Give **one** example of a human disease caused by a point mutation. **[1]**

(*c*) Scientists have determined that there are 56 differences between the mitochondrial DNA of chimpanzees and humans. The number of differences between Neanderthal man and modern humans is 28. There are 7 differences between the different races of modern humans in the world today.

Using this information, calculate how long ago modern humans split from

(i) Chimpanzees

(ii) Neanderthal man **[2]**

(*d*) The diagrams below show two possible evolutionary relationships between chimpanzees, Neanderthal man and modern humans.

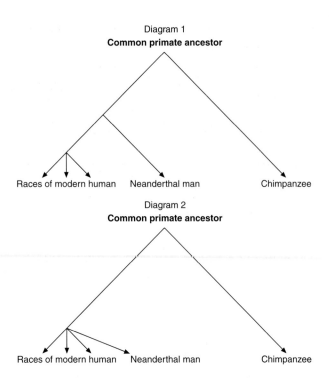

Which of the evolutionary relationships is best supported by the evidence from mitochondrial DNA? Explain your answer. **[3]**

(*e*) The use of information from mitochondrial DNA is just one method of determining the evolutionary relationship between hominoids. Describe **one** other technique that does **not** involve DNA, which can be used to estimate the time at which evolutionary lines of two hominoids diverged. **[3]**

(*f*) Many complete skeletons of Neanderthal man have been discovered in Europe. Some of the skeletons have been found in caves together with the remains of animals skins, animal bones and pollen grains. What can be deduced about Neanderthal culture from this information? **[3]**

(Total 14 marks)

(Edexcel GCE Biology (Human) Unit Test 6115/01, June 2002)

3 The graph below shows the mean brain size in four species of hominids.

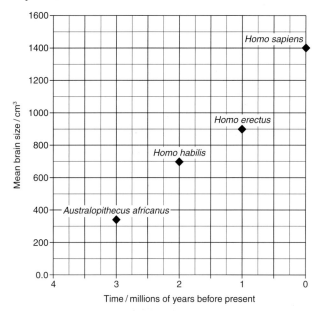

(a) Using the information in the graph, describe the changes in mean brain size over the last three million years. **[2]**

(b) State **two** changes that occurred in the hominid skeleton over this period of time. **[2]**

(c) Describe how the age of an unknown hominid fossil that is estimated to be several million years old could be determined. **[4]**

(Total 8 marks)

(Edexcel GCE Biology (Human) Unit Test 6115/01, June 2003)

4 The graph below shows human survival patterns in two populations, the USA and India, at the beginning of the 20th century.

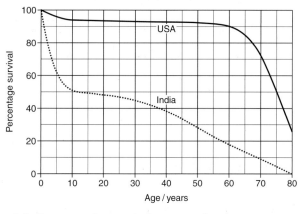

(a) Compare the survival patterns for the two populations. **[3]**

(b) With reference to the information provided, suggest reasons for the differences in the survival patterns. **[3]**

(Total 6 marks)

(Edexcel GCE Biology (Human) Unit Test 6115/01, June 2003)

5 (a) Complete the table below by writing in the name of the species of *Homo* in which the features first appeared. **[3]**

Features	Species of *Homo*
Bipedalism and toolmaking	
Culture and religious beliefs	
Use of fire	

(b) Describe **two** advantages of bipedalism. **[2]**

(Total 5 marks)

(Edexcel GCE Biology (Human) Unit Test 6115/01, January 2003)

Mark schemes

In the mark schemes the following symbols are used:
; indicates separate marking points
/ indicates alternative marking points
eq. means correct equivalent points are accepted
{} indicate a list of alternatives

Chapter 1

1 (*a*) It is a factor which limits the rate (of photosynthesis / of a reaction) ; Increasing this factor increases the rate (of photosynthesis) ; Until another factor limits the rate / eq ; **[2]**

    (*b*) (i)
1. Increase in rate from 0 to 300 / 350 (light intensity) ;
2. Reference to manipulated figures to support increase ;
3. Light is the limiting factor / eq ; [accept converse]
4. Rate levels off / eq above 300 / 350 (light intensity) ;
5. Another factor has become limiting ;
6. Named limiting factor: carbon dioxide / temperature / chloroplasts / chlorophyll ; **[4]**

        (ii)
1. (Provides) carbon dioxide ;
2. Rate of, O_2 production / photosynthesis, is greater (at all light intensities) above 50 for higher carbonate / CO_2 concentration / eq ;
3. Reference to comparative figures for rates / approx. double ;
4. For both concentrations rate starts to level off at 300 / 350 light intensity / lower concentration levels off above 250 compared with 300 ;
5. Hydrogencarbonate / CO_2 concentration is a limiting factor / not limiting up to light intensity 50 ; **[3]**

        (iii) Keep temperature constant / place tube containing pond weed in a water bath ; Reference to rate of reaction / enzymes being affected by temperature changes / temperature affects, solubility / diffusion / volume of oxygen / kinetic energy ;

Repeat / replicate ;
To avoid anomalous / eq results ;
Use the same piece / length of pond weed throughout ;
A different piece could have, more / fewer, leaves / different surface area / eq (and so affect the volume of oxygen produced) ;
Reference to equilibration / eq ;
To make sure pond weed is acclimatised to new conditions ;
No air bubbles in equipment ;
So measuring only gas / oxygen given off ;
Same type of light source ;
So keep wavelength the same ;
No external light sources ;
To control the experimental light intensities ; **[1+1]**

(*c*)
1. Splitting water in (presence of) light ;
2. $H_2O \rightarrow \frac{1}{2}O_2 + 2e^- + 2H^+$; [allow words]
3. Electrons replace the electrons lost from PS11 / P680 / chlorophyll ;
4. Hydrogen ions picked up by NADP / hydrogen ions reduce NADP ;
5. NADPH / eq used, in light-independent reaction / Calvin cycle / to reduce carbon dioxide ; **[3]**
                                     (Total 14 marks)

2 (*a*) Affects (rate of) enzyme activity / affects (rate of) enzyme {controlled / catalysed} reaction / correct reference to kinetic energy of molecules / reactants / eq ;

In light-independent stage / Calvin cycle ; **[2]**

    (*b*)
A Starch (grain) ;
B Stroma ;
C Granum / stack of thylakoids ;

[All correct for two marks, two correct for one mark] **[2]**

    (*c*) (i) X CO_2 / carbon dioxide ;
             Y PGA / phosphoglyceric acid /

phosphoglycerate / GP / glycerate 3-phosphate / unstable 6-carbon compound ; **[2]**

   (ii) Light-dependent reaction / grana / thylakoid / non-cyclic photophosphorylation ; **[1]**

(*d*) 1. Sucrose ;

   2. Loading by {companion cells / transfer cells / loading cells} ;

   3. Active process / requires {energy / ATP} ;

   4. In phloem sieve {tubes / elements} ;

   5. Reference to {reduction in cell organelles / sieve plates / perforated end walls} ;

   6. Continuous tube ;

   7. Mass flow ;

   8. Bi-directional flow / flow in any direction ;

   9. From {leaves / eq} to {growing regions / storage organs} / reference to source to sink ;

   10. Oxygen by diffusion ; **[5]**
(Total 12 marks)

3 (*a*) Two peaks / eq at 435 – 440 and 665 – 675 nm ;

Little absorption // eq between 450 and 600 // 650 nm ;

Second peak smaller / lower than the first / quoted difference / no absorption 470 – 480 nm ; **[2]**

(*b*) Each (pigment) absorbs a different wavelength of light ;

(More pigments) gives a greater range of absorption / plants able to absorb light of greater range of wavelengths ;

Reference to protective nature of carotenoids / accessory pigment qualified ; **[2]**

(*c*) Use chromatography ;

Grind up / crush sample of plant tissue in propanone / organic solvent / eq ;

Place drops of pigment / eq on paper / gel / column / silica ;

Suspend paper in test tube / boiling tube with bottom touching solvent ;

Time / eq until pigments are seen to separate ; **[4]**
(Total 8 marks)

Chapter 2

1

Plant growth substance	One function or effect
	Promotes – cell elongation / apical dominance / root formation in cuttings / lateral roots / fruit development OR inhibits – lateral bud development / abscission of {leaf / fruit} ;
Cytokinin ;	
	Promotes – seed germination / breaking of seed dormancy / cell elongation / stem elongation / bolting in {long day plants / rosette plants / dwarf plants} / fruit development / seedless fruit ;
Ethene / C_2H_4 / ethylene ;	
	Promotes – {leaf / fruit} fall / leaf abscission / stomata closure / seed dormancy / bud dormancy / stress proteins OR Inhibits – cell division / growth / germination ;

(Total 5 marks)

Chapters 3, 4 and 5

1

Kingdom	Animalia
Phylum ;	
Order ;	
	Pan ;
Species	

(Total 4 marks)

MARK SCHEMES

2 (*a*) Biomass / energy transferred per unit area per unit time ;

NPP = GPP–R / Gross production minus respiration ;

Biomass / energy that is available to primary consumers / herbivores / next trophic level ; **[2]**

(*b*) (i) (Abandoned land) colonised by weeds / ruderals / grass plants / grasslands ;

(Grassland) replaced by (small) shrubs / scrub / eq ;

(Scrub) replaced by woodland / trees / woods / eq ;

Climax (woodland / forest) ;

Increase in biodiversity / increase in number of species / increase in habitat diversity ; **[3]**

(ii) 150 (%) ; **[1]**

(iii) Same biomass may have different energy content ;

Example ;

Energy content indicates what is available to next trophic level / eq / reference to energy flow / energy losses along food chains / between trophic levels ; **[2]**

(iv) Grass smaller than trees so biomass (per m²) is lower (than that of trees) ;

As plants get larger / grasses replaced by trees, biomass increases ;

Grasses produce a lot of edible material / large proportion of the net primary productivity of grasses passes to consumers / herbivores ;

Trees produce wood / tree biomass accumulates ;

NPP levels off after 40 (to 60) years as climax community is reached / eq ;

Reference to a factor which limits further increase in NPP (e.g. light) ; **[3]**

(Total 11 marks)

3 (*a*) (A chemical substance that) does not break down naturally / is not broken down by living organisms / eq ;

Persists in the environment ; **[2]**

(*b*) Pesticide taken up from the water into the plankton ;

(Plankton) eaten by fish ;

Pesticide stored in body (fat of fish) ;

Grebes eat a large number of fish (containing DDT) ;

Reference to bioaccumulation along food chain / eq ; **[3]**

(*c*) Biological control ;

Introduce a predator / parasite to the midge ;

Always have a small population of midges / midges not completely wiped out ;

Introduce sterile males ;

Reduce successful reproduction / eq ; **[2]**

(Total 7 marks)

4 (*a*) (i) Use of one organism / eq to control / eq another organism / pest ;

By feeding on it / predation / parasitism / causing disease ; **[2]**

(ii) Feeding / eq on another living organism ;

And causing harm / eq ; **[2]**

(*b*) (i) Whitefly population / eq grows faster (than wasp population) ;

So predation (rate) is lower / eq / not enough wasps to control / eq whitefly ; **[2]**

(ii) Wasps may (nearly) eliminate the pest / whitefly / eq ;

(So) wasps die out / wasp population falls ;

Pest / whitefly could return / eq ; **[2]**

(Total 8 marks)

Chapter 6

1 (*a*) (i) Normal distribution (curve) / no distinct groups ;

Polygenes / produced by (combined) effects of many {genes / loci} ;

Reference to (significant) environmental influences ;

Example (body mass, shape, intelligence, skin colour, height) ; **[2]**

(ii) Distinct difference in characteristic / no

intermediates / discrete groups ;

{Few / two / single} gene with two or more alleles ;

No environmental influence ;

Example (tongue rolling, sex, eye colour, ear lobe, hairy ears, genetic diseases, blood groups) ;   **[2]**

(*b*) (i)   Substitution of A for T ;

{Addition / insertion} of A ;

Deletion of T if next codon base is A ;

Inversion if next codon base is A ;   **[2]**

(ii) 1.   {Degenerate / redundant} code OR only first two bases necessary to code for a particular amino acid / eq ;

2.   (if substitution) only alters one amino acid ;

3.   (therefore) no significant change in protein structure ;

4.   So no change in {protein / enzyme} function ;

5.   (maybe) no frame shift ;

6.   Could be in {non-functional DNA / junk DNA / intron} ;

7.   Allele could be recessive ;   **[2]**

(*c*)

Gametes	Ab	ab
aB	AaBb normal	aaBb deaf
ab	Aabb deaf	aabb deaf

Gamete correct ;

Correct genotype ;

Correct phenotype linked to genotype ;

Correct calculation (1 in 4 / 25% / 0.25) ;

[If genotypes wrong, allow correct phenotypes and calculation for max 2 marks]   **[4]**

(Total 12 marks)

2 (*a*)   More than two (alleles) ;

(Alternative) forms / eq of a gene ;

At same, locus / position *or* on homologous chromosomes ;   **[2]**

(*b*) (i) I^AI^B and I^BI^O ;   **[1]**

(ii)   Gametes correct (accept from diagram) ;

I^AI^B, I^OI^O, I^AI^O and I^BI^O (are the offspring genotypes) ;

A, B, AB or O (offspring phenotypes) ;   **[3]**

(Total 6 marks)

Chapter 7

1   Your essay should contain references to the following:

Outline structure and functions of male and female reproductive systems –

Structure of gametes –

Transfer of gametes –

Features of sperm in relation to fertilisation –

Fusion of male and female nuclei to produce a diploid zygote –

Genetic screening and detection of fetal abnormalities by aminocentesis and chorionic villus sampling –

Karyotypes –

Possible course of action –

Implications (social, ethical and legal) –

Scientific content [13]

Balance [2]

Coherence [2]

(Total maximum 15 marks)

Chapters 8 and 9

1 (*a*)   Small ;

Circular / eq DNA ;

Found in prokaryotes ;

Separate from bacterial DNA / main chromosome / replicates independently ;

Carries, resistance / sex, genes ;

Can be transmitted from one bacterium to another ;   **[3]**

(*b*)   To show / identify the (herbicide resistant) gene ;

Has been incorporated into vector / bacterium / (host) cell ;

Example of a marker gene ;   **[2]**

MARK SCHEMES

(c) 1. RNA polymerase ;

2. Attaches to DNA in region of gene / at start signal ;

3. H bonds broken / DNA unwinds / eq / separates into two strands ;

4. One of the DNA strands ;

5. Reference to complementary base pairing (of nucleotides) ;

6. New DNA base sequence copied into mRNA / eq ;

7. Reference to, introns removed / adenine tail / guanine cap, added ; **[4]**

(d) Resistance may spread to other plants / hybridisation with other plants ;

Plants become, invasive / difficult to control ;

Modified / eq plants may become, toxic / allergenic, to consumers ;

(therefore) Use may increase disease transmission ;

Vector is a plant pathogen ;

Reference to contamination of non-GM crops / seeds / organic farming / consumer resistance ; **[3]**

(Total 12 marks)

2 (a) (Organism) whose DNA / genome / genetic code / genetic sequence / genotype / gene has been altered ;

Artificially by gene technology / genetic engineering ;

Reference to transgenic organism containing foreign DNA / recombinant DNA / recombinant gene ; **[3]**

(b) (i) Reverse transcriptase ; **[1]**

(ii) Complementary / matching base pairing (of nucleotides) ;

On cDNA single strand ;

Forms DNA double strand ;

(DNA) polymerase ; **[3]**

(iii) Coagulation / clotting of milk / casein / cheese making / curds making ; **[1]**

(iv) Suitable for vegetarian diets / does not involve killing animals ;

Can link supply and demand closer / faster / greater production ;

Less risk of contamination / eq ; **[2]**

(Total 10 marks)

Chapters 10 and 11

1

Group	External features
Apes	No tail ; (Nose) downward pointing / narrow nostrils ; Arms swing freely / can swing arms over head / eq ; Nails ; Opposable thumb / digit ;
New world monkeys	Prehensile tail ; Broad / flat / wide nose / widely separated nostrils ; Nails ;

(Total 2 + 2 marks)

2 (a) Change in a single base / one base / nucleotide ;

By deletion / addition / substitution / insertion ;

Causes change in amino acid sequence / changes amino acid coded for ; **[2]**

(b) Sickle cell anaemia (or other correct answer) ; **[1]**

(c) (i) (56 × 21 500 =) 1 204 000 years ;

(ii) (28 × 21 500 =) 602 000 years ; **[2]**

(d) Diagram 1 ;

More / 28 differences between Neanderthal man and modern man, fewer / 7 differences between different races of man ;

Neanderthals diverged before the different races of humans diverged / about half way between modern humans and chimps / reference to the fact that the divergence between chimps and modern man was twice the time of the divergence between Neanderthal man and modern man / appropriate use of figures / eq ;

OR

In diagram 2 Neanderthals and human races shown as diverging at the same time ; **[3]**

(e) **Any one of the following methods:**

Antigen / antibody / immunoglobulin test / reaction ;

Serum / antigens from primate / human injected into animal / rabbit, obtain (sensitised) rabbit serum / antibodies ;

React / mix rabbit serum / antibodies with serum / antigens from other animals ;

Measure quantity of precipitate / eq ;

Amino acid sequence ;

Determine amino acid sequences in proteins / haemoglobin ;

Compare to sequences of same protein in other animals ;

The more in common / similar, the more closely related the animals ;

Radiometric / eq method ;

Based on the rate at which (radioactive) isotopes decay / eq ;

Calculation of age of sample based on the half life of isotope ;

Measure the proportion of the particular isotope present in the sample ;

Geochronology / lithostratography ;

Examine fossils ;

From same strata / same layers ;

Compare with fossils of known age / rock evidence ;   **[3]**

(f) Neanderthals buried their dead ;

Could be a religious ritual ;

Food / plants / flowers / eq buried with body ;

Used animal skins for clothing / shelter ;

Hunted / ate other animals / hunter-gatherers ;

Mixed diet ;

Not nomadic / converse ;

Use of tools (to skin animals) ;   **[3]**

(Total 14 marks)

3 (a) Increase in brain size almost linear / steady increase ;

Overall $\times$ 4 increase in brain size / increase in mean brain size of 1050 / 1060 cm^3 ;   **[2]**

(b) Change in jaw shape to become rounded / smaller / less prominent / teeth smaller ;

Reduced brow ridge ;

Increase in height ;

Arms relatively shorter (relative) to height / legs / reference to reduction in opposability in feet ;

Increase in size of cranium ;

Reference to changes in pelvic girdle / hips for upright posture / eq ;   **[2]**

(c) **Radiometric method:**

Use radiometric dating / use radioactive isotopes ;

Potassium–argon dating (as several million years old) ;

Based on decay of (radioactive) isotopes ;

Reference to half life ;

Ratio of isotopes / ratio of ^{40}K : ^{40}Ar gives estimate of age ;

OR geochronology method:

Use geochronology / lithostratography ;

Rock / material found around the fossils is studied / study the sedimentary layers above / below the fossil ;

Can date the rocks using radiometric / K–Ar methods ;

Compare with other (known) fossils found in the same layer of rock ;

Study any artefacts / objects used by humans / tools found with the fossil ;   **[4]**

(Total 8 marks)

4 (a) USA survival rates higher at all ages / converse ;

Survival rate at 8–10 years much greater in USA than in India ;

Little change in USA between 8–10 and 58–61 years and steady decline in India ;

Rapid decline between 58–61 and 80 years in USA and in India no survivors / eq at 80 years ;

Credit comparative manipulation of figures ;   **[3]**

(b) Better control of child related diseases / vaccinations in USA / converse ;

MARK SCHEMES

Cleaner water supply in USA / better sanitation / hygiene / housing / diet / healthcare in USA / converse ;

(Decline in USA) due to diabetes / coronary heart disease / stroke / cancer / eq ; **[3]**

(Total 6 marks)

5 (*a*)

Features	Species of *Homo*
	habilis ;
	sapiens ;
	erectus ;

[3]

(*b*) Frees hands for using tools / using weapons / holding babies / manipulation ;

Taller so sees further ;

Longer gait, faster movement ; **[2]**

(Total 5 marks)

Index

Page references in italics refer to a table or an illustration.

INDEX

INDEX